POLYMER SCIENCE AND TECHNOLOGY
Volume 7

BIOMEDICAL APPLICATIONS OF POLYMERS

POLYMER SCIENCE AND TECHNOLOGY

Volume 1 • STRUCTURE AND PROPERTIES OF POLYMER FILMS
Edited by Robert W. Lenz and Richard S. Stein • 1972

Volume 2 • WATER-SOLUBLE POLYMERS
Edited by N. M. Bikales • 1973

Volume 3 • POLYMERS AND ECOLOGICAL PROBLEMS
Edited by James Guillet • 1973

Volume 4 • RECENT ADVANCES IN POLYMER BLENDS, GRAFTS, AND BLOCKS
Edited by L. H. Sperling • 1974

Volume 5 • ADVANCES IN POLYMER FRICTION AND WEAR (Parts A and B)
Edited by Lieng-Huang Lee • 1974

Volume 6 • PERMEABILITY OF PLASTIC FILMS AND COATINGS TO GASES, VAPORS, AND LIQUIDS
Edited by Harold B. Hopfenberg • 1974

Volume 7 • BIOMEDICAL APPLICATIONS OF POLYMERS
Edited by Harry P. Gregor • 1975

POLYMER SCIENCE AND TECHNOLOGY
Volume 7

BIOMEDICAL APPLICATIONS OF POLYMERS

Edited by
Harry P. Gregor
Department of Chemical Engineering and Applied Chemistry
School of Engineering and Applied Science
Columbia University
New York, New York

PLENUM PRESS · NEW YORK AND LONDON

Library of Congress Cataloging in Publication Data

Main entry under title:

Biomedical applications of polymers.

(Polymer science and technology; v. 7)
"A symposium held by the American Chemical Society, Division of Organic Coatings and Plastics Chemistry, in Chicago, Illinois, in August, 1973."
Includes bibliographical references and index.
1. Polymers in medicine–Congresses. 2. Biomedical materials–Congresses. I. Gregor, Harry P. II. American Chemical Society. Division of Organic Coatings and Plastics Chemistry. III. Series. [DNLM: 1. Biomedical Engineering–Congresses. 2. Polymers–Congresses. QT34 A514b 1973]
R857.P6B56 610'.28 75-6846
ISBN 0-306-36407-7

Proceedings of a symposium held by the American Chemical Society,
Division of Organic Coatings and Plastics Chemistry,
in Chicago, Illinois, in August, 1973.

A Division of Plenum Publishing Corporation
227 West 17th Street, New York, N.Y. 10011

United Kingdom edition published by Plenum Press, London
A Division of Plenum Publishing Company, Ltd.
Davis House (4th Floor), 8 Scrubs Lane, Harlesden, London, NW10 6SE, England

Printed in the United States of America

This volume is dedicated to

Maurice B. Visscher, M. D.

Professor Emeritus of Physiology, The Medical School,
University of Minnesota, a pioneer in interdisciplinary
studies which now constitute the field of Bioengineering

Foreword

The intense integration of physical and biological science that has developed in the last several decades has led to major problems in the wise use of available knowledge, both conceptual and factual. In the last century, before the separation of the physical and biological sciences became distinct, all the knowledge recognizable as bearing on a biological problem was relatively small and one man, albeit often an exceptional one, could contribute importantly to the advance of that knowledge.

Now the recognizable knowledge pertaining to living systems is much larger, our questions are correspondingly more complex, and our goals are set higher. As many of the papers in this symposium show, we aim beyond analyzing injury and disease toward the synthetic goals of imitating arrested or inadequate life processes and reconstructing maimed organisms. Such activity supposes adequate knowledge of the natural functions to be replaced, and, because artificial materials are seldom indistinguishable from their natural archetypes, the prosthetic attempt raises a host of new questions about the reactions between artificial and living materials in intimate contact.

Among problems of this sort are: thromboresistance, the quality by which an artificial surface avoids activating clotting enzymes or allowing cells to stick or be activated for thrombus formation; tissue compatibility; maintenance of chemical and mechanical integrity of artificial materials in contact with body fluids; and absence of tissue irritation. All of these desiderata have been considered here, some from several points of view.

I am especially familiar with the thromboresistance problem, which has turned out to be surprisingly difficult, at the level both of its amelioration and of its understanding. Duplicating the passivity of vascular endothelium to blood seemed at first to be an easy task. It soon became clear that understanding of the endothelial-blood symbiosis was complex and inadequate to the specification or development of artificial thromboresistant materials. Supposed solutions to the problem fell short of what was desired. Some of these attempts raised their own problems whose analysis represented a diversion of effort from the main goal, but others placed fundamental problems in a new light. Thus, some work on thromboresistance (including much of that reported here) contributed not only to the improvement of thromboresistant materials but also to a better understanding of the mechanism of thrombosis *in vivo* and various coagulopathies. The synthetic goal stimulated analysis of normal and abnormal function, and it brought new disciplines into contact with a fundamental biological problem.

Chemistry has played a special role in sustaining some contact between medical and biological science and physical science. Sight was never lost of the fundamental chemical nature of life processes. Biochemistry was one of the earliest, most heavily populated of the bio-something sciences. Still, the exchange of knowledge and overlap of education between the industrially oriented and the biochemically oriented chemist was seriously limited. A major conceptual difference resulted insofar as the former group was concerned primarily with synthesis while the latter was largely concerned with the analysis of pre-existing, highly complex systems.

A number of presentations here deal with model systems whose utility may reach to both an understanding of natural processes and a basis for exploiting natural processes for new purposes or for replacing them to compensate for disease or injury. The synergistic effect of contiguous efforts at synthesis and analysis is great even when final benefits are not completely responsive to original goals.

This timely symposium, therefore, strikes me as rich, not only for its specific content and its testimony of solid technical progress, but also because it shows a new confluence of thought between the analyzers of natural chemical processes and materials and the synthesizers of artificial chemical processes and materials which augment, replace, and emulate nature. The range of problems considered here precludes a full-faceted, definitive treatment of any one, but it shows a healthy catholicity of thought and some surprising commonality of approach to diverse problems of creating compatible and synergistic juxtapositions of artificial and living materials.

E. F. Leonard

Columbia University
December, 1974

Contents

THE HYDRATION OF PHOSPHOLIPID FILMS AND ITS RELATIONSHIP TO PHOSPHOLIPID STRUCTURE

J. H. Hasty* and G. L. Jendrasiak

Biophysics Division and
Program in Bioengineering
University of Illinois
Urbana, Illinois 61801

Little work has been done on the hydration of lipids and its concomitant effect on their electrical conductivity. Since the membrane-water interface is thought to play an important part in membrane permeability, such studies may have biological relevance. The nature of the water bound by the lipids might be expected to play some part in the structure of the membrane interface region in membranes containing these lipids. Elworthy (1,2) has obtained the adsorption isotherms for certain phospholipids, whereas Jendrasiak (3) has investigated the hydration and concomitant electrical conductivity of egg phosphatidylcholine. The purpose of the work reported in this paper is to determine the adsorption isotherms for egg phosphatidylcholine (PC), egg phosphatidylethanolamine (PE), and bovine phosphatidylserine (PS), in both their lyso and diacyl forms. The effect of complexation of PC with cholesterol and the effect of varying the number of double bonds in the hydrocarbon chains are also studied. In this way, it is hoped to obtain some idea as to the relative contribution to hydration of these lipids by the hydrocarbon chains and polar head groups, respectively. Also, the effect of varying the nature of the polar head group on lipid hydration can be studied. The effect of the hydration on electrical conductivity of the lipids is also studied. The complete study will be published (4).

---

*Kendall Research Center, Barrington, Ill., 60010

MATERIALS AND METHODS

Lipid films were cast, from the appropriate lipid dissolved in chloroform, on thin teflon strips. After removal of the solvent the films were placed in a Cahn G-2 electrobalance operating in the "remote weighing" mode. The entire assembly was separated from the balance controls and placed in a chamber where both the relative vapor pressure (P/Po) and temperature (22 ± 0.15°C) were controlled. The controlled humidities were obtained by means of saturated salt solutions. The electrical conductivities of the films were measured using films cast on quartz and an electrometer circuit.

RESULTS AND DISCUSSION

From Fig. 1 it can be seen that egg PC, in its diacyl form, exhibits, according to Brunauer's (5) classification scheme, a type II or IV isotherm. Type II isotherms are frequently encountered and represent multilayer physical adsorption by non-porous solids or microporous solids. From the inflection points of such a plot, the amount of water necessary for the formation of an adsorbed monolayer can be obtained: We judge this value to be near 2.0 water molecules adsorbed per lipid molecule. Our scanning electron micrographs do reveal some structure in the egg PC films which might represent a certain degree of porosity.

Egg PE and bovine PS, in their diacyl forms display type III or V isotherms. Such isotherms are indicative of no rapid initial uptake of water vapor and occur when the forces of adsorption in the first monolayer are relatively small.

All of the phospholipids, in their lyso form display type III (or V) isotherms, again suggesting no rapid initial uptake of water, as seen in Fig. 2.

Type IV and V isotherms exhibit saturation which apparently reflects capillary condensation.

Since the isotherms for the diacyl and lyso form of PC differ in shape, it appears that removal of a hydrocarbon chain in PC does have a significant effect on the water adsorption characteristics. In the case of PE, on the other hand, the isotherms for the lipid in both the

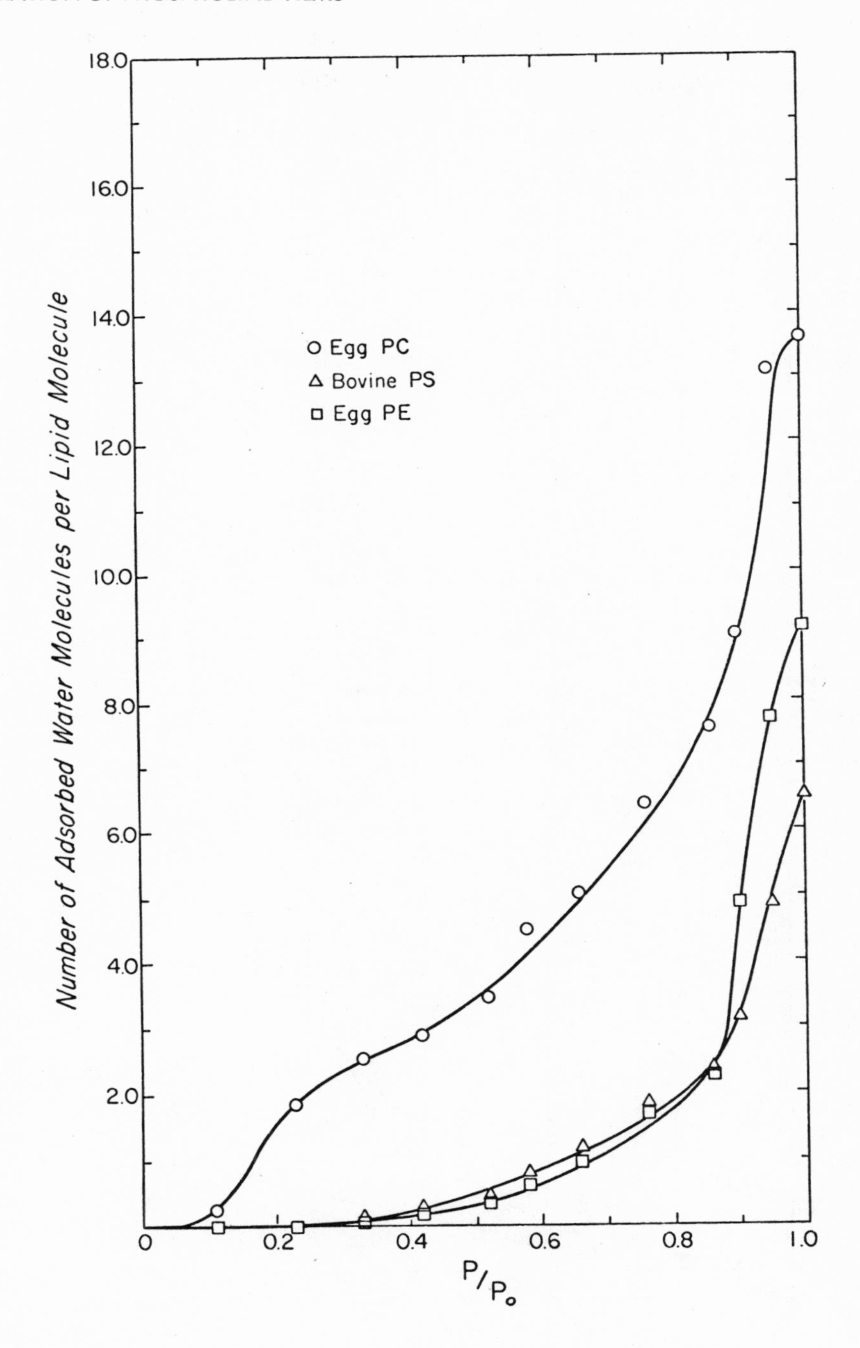
18.0
16.0
14.0
12.0
10.0
8.0
6.0
4.0
2.0
Number of Adsorbed Water Molecules per Lipid Molecule
○ Egg PC
△ Bovine PS
□ Egg PE
0
0.2
0.4
0.6
0.8
1.0
P/Po

Fig. 1

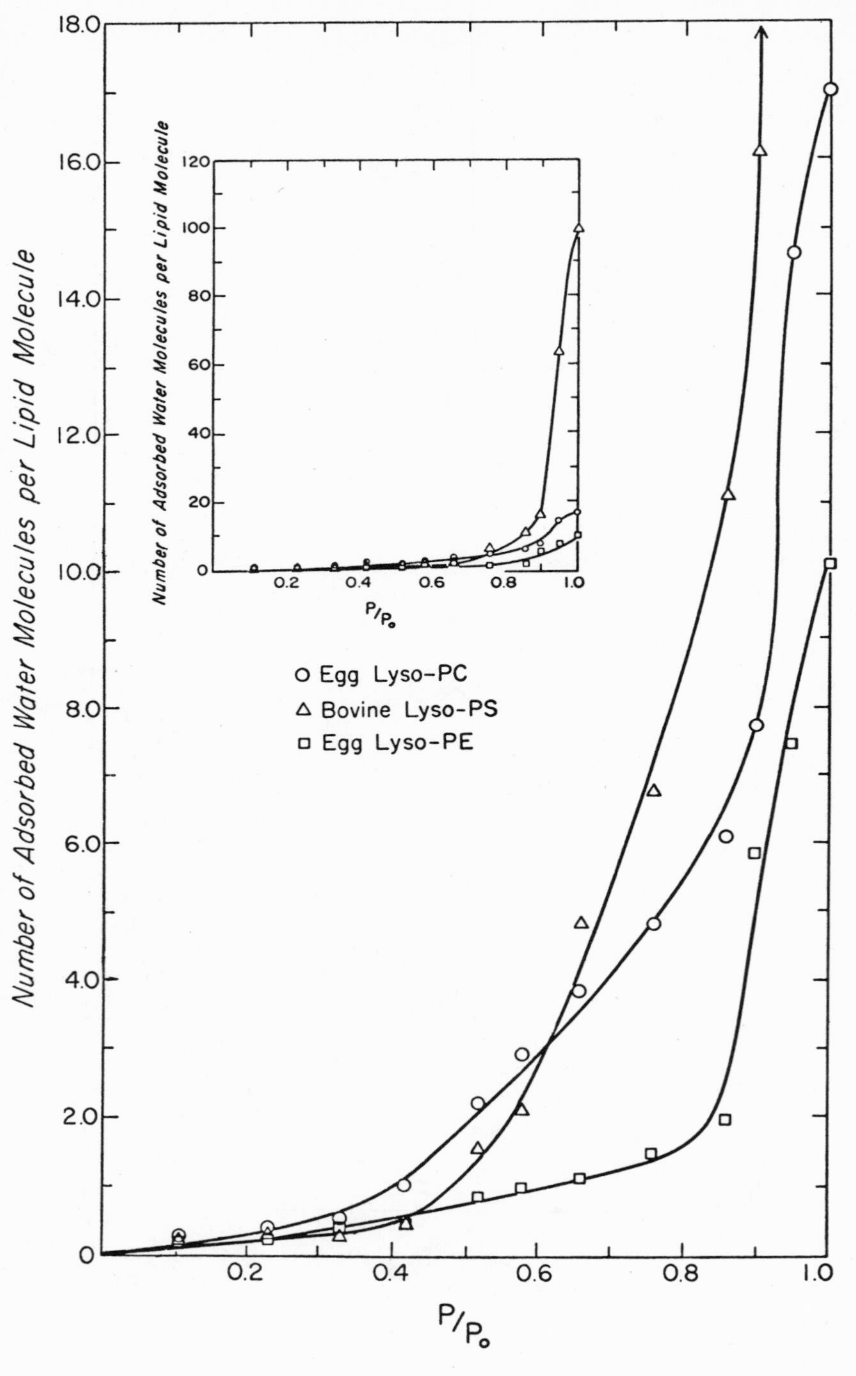
Number of Adsorbed Water Molecules per Lipid Molecule
18.0
16.0
14.0
12.0
10.0
8.0
6.0
4.0
2.0
0
0.2
0.4
0.6
0.8
1.0
P/P₀
Egg Lyso-PC
Bovine Lyso-PS
Egg Lyso-PE
120
100
80
60
40
20
0

Fig. 2

diacyl and lyso forms are almost identical; this suggests that the polar head group plays the dominant role in the water adsorption behavior of this lipid. The lyso form of PS adsorbs more water than does the diacyl form at all vapor pressures. Particularly striking is the fact that lyso-PS adsorbs some one hundred molecules of water per molecule of lipid at a vapor pressure of 1.0, whereas for the diacyl form of PS, the figure is closer to six. This suggests that the removal of a hydrocarbon chain has a very strong influence on the water adsorption of PS. This deviation of the lyso-PS isotherm from the diacyl PS isotherm does not become apparent until the lyso-PS has adsorbed about one water molecule per two molecules of lipid. It may well be that at this amount of water, the polar head group of the serine is ionized and the sample micellizes with ionic regions in between the micelles.

The effect of the complexation of cholesterol with egg PC (1:1,mole:mole) is shown in Fig. 3. The isotherm displayed by the complex is a type II (or IV) suggesting that the cholesterol molecule has no strong effect on the configuration of the polar head group of PC. The amount of water adsorbed per PC molecule is increased, over the entire vapor pressure range, from the corresponding values with no cholesterol present. Since the cholesterol itself adsorbs a negligible amount of water compared to PC, the increase may be due to some effect on the PC by the cholesterol. We postulate that the effect is due to an increase in the size of the cavity occupied by the PC head group, allowing more water to enter the cavity. It has been found by other methods that complexing cholesterol to PC increases the water bound to PC; our results support such a conclusion.

Fig. 4 illustrates the effect of increasing the number of double bonds in the hydrocarbon chains of diacyl PC, on the water adsorption characteristics of the lipid. Increasing unsaturation increases the amount of water adsorbed at all vapor pressures. There is some question as to whether the increase in water adsorption with unsaturation arises because of double bonds in both hydrocarbon chains, or in only one of the hydrocarbon chains. From Fig. 4, one can see that the basic character of the isotherm is established by the polar headgroup with the hydrocarbon chains having a strong modulating effect on the isotherms. All of the isotherms remain type II (or IV), suggesting that the head group

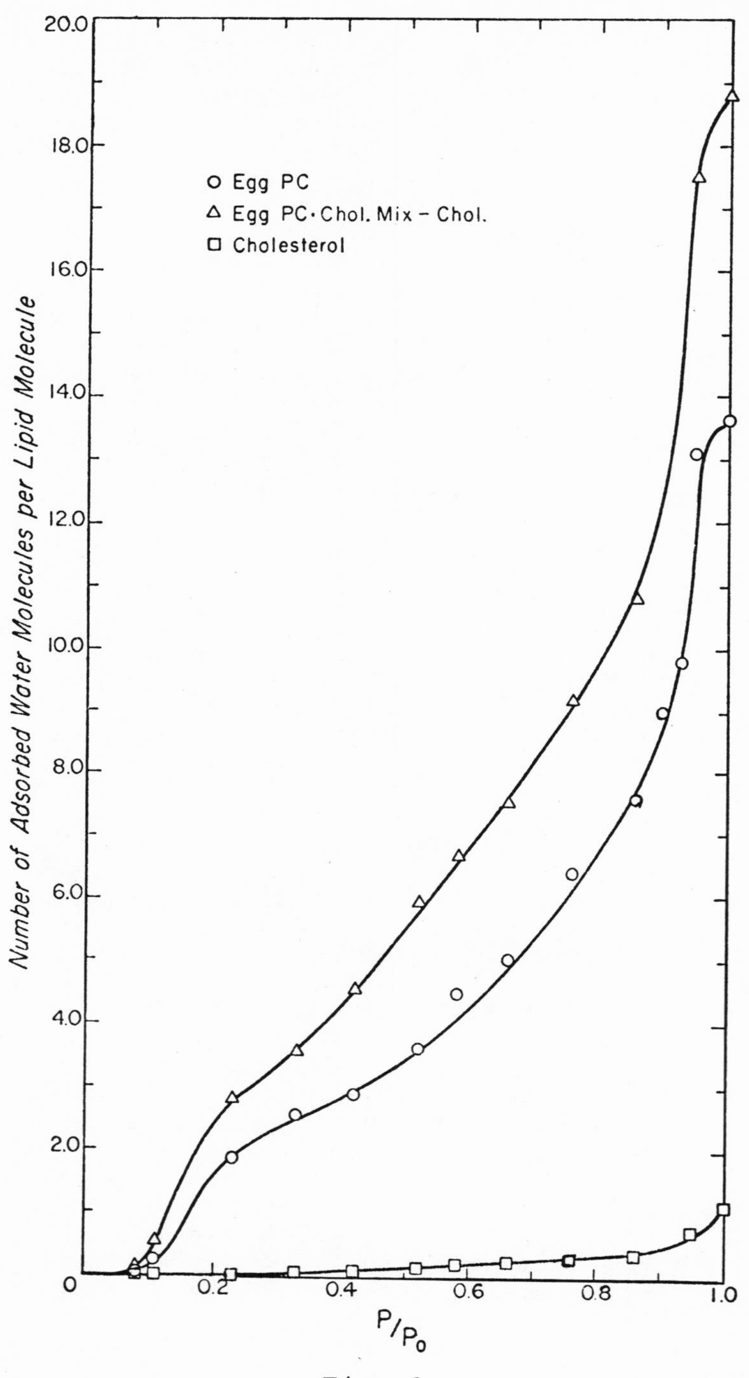

20.0
18.0
16.0
14.0
12.0
10.0
8.0
6.0
4.0
2.0
0
0.2
0.4
0.6
0.8
1.0
Number of Adsorbed Water Molecules per Lipid Molecule
P/P0
Egg PC
Egg PC·Chol. Mix – Chol.
Cholesterol

Fig. 3

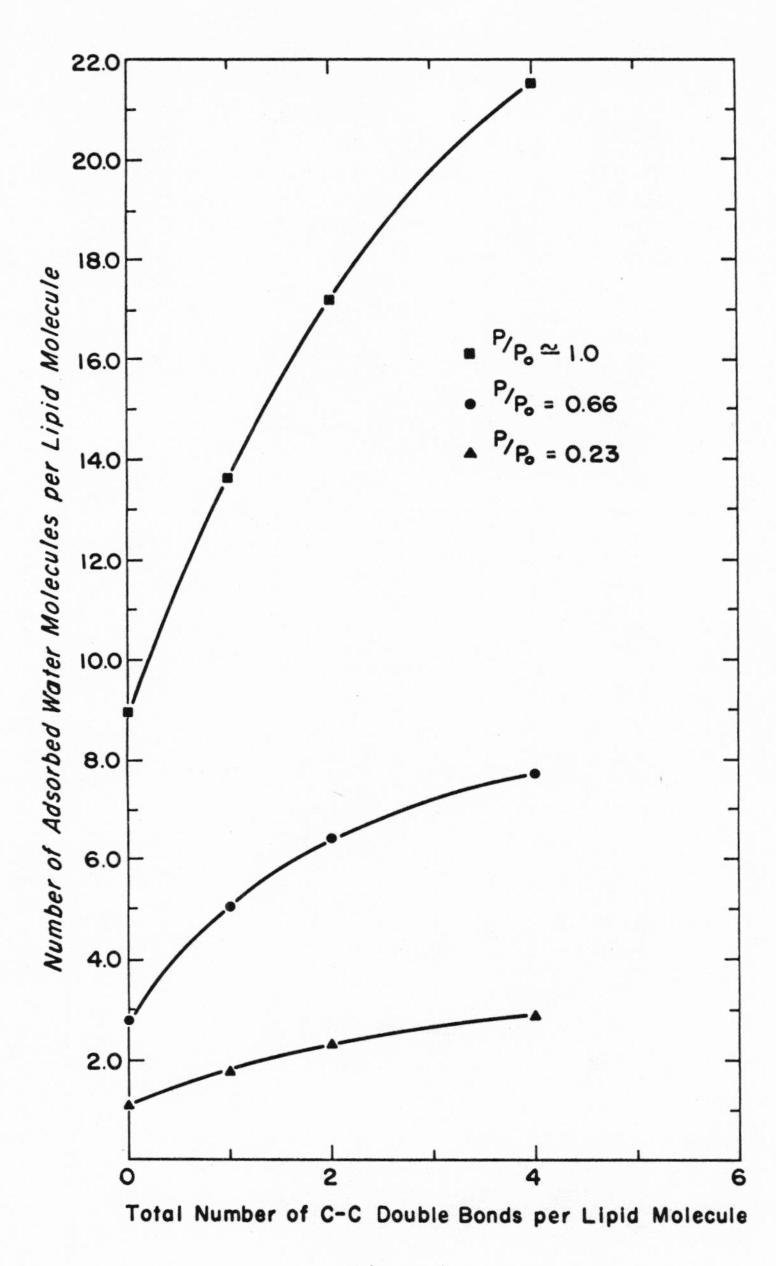
22.0
20.0
18.0
16.0
14.0
12.0
10.0
8.0
6.0
4.0
2.0
0
2
4
6
Number of Adsorbed Water Molecules per Lipid Molecule
Total Number of C-C Double Bonds per Lipid Molecule
P/P₀ ≃ 1.0
P/P₀ = 0.66
P/P₀ = 0.23

Fig. 4

has a strong effect on the water adsorption character of PC.

It should be noted that diacyl egg PE apparently belongs to a different isotherm class than does egg PC. The absence of the CH_3 groups on the nitrogen of egg PE seemingly alters the polar head group in such a way as to allow fewer water molecules to be adsorbed in the PE head-group region than for the PC headgroup region. This may come about because of a possible extended conformation of the zwitterionic head group in PC, whereas in PE the extended conformation may be absent. (6)

Other workers (7) have studied lamellar phases of lipid-water systems by X-ray techniques. Our results based on PS and cardiolipin suggest that the X-ray results and our results may not be directly comparable.

The electrical conductivities of all of the lipid films have been measured as a function of vapor pressure. Also, the activation energies for electrical conductivity, at $P/P_O = 1.0$, have been measured for all of the lipids.

ACKNOWLEDGMENTS

One of us (JHH) was supported by USPHS grant GM720. This work was also partially supported by funds made available by the University of Illinois Research Board.

REFERENCES

1. Elworthy, P. H., J. Chem. Soc. (1961) 5385.
2. Elworthy, P. H., J. Chem. Soc. (1962) 4897.
3. Jendrasiak, G. L., Ph.D. Thesis, Michigan State University (1967).
4. Jendrasiak, G. L. and Hasty, J. H., Biochim. Biophys. Acta (In press)
5. Brunauer, S., The Adsorption of Gases and Vapors, Vol. 1, Princeton University, Princeton, New Jersey, (1945) p. 150.
6. Phillips, M. C., Finer, E. G. and Hauser, H., Biochim. Biophys. Acta, 290 (1972) 397.
7. Gulik-Krzywicki, T., Tardieu, A. and Luzzati, V., Molecular Crystals and Liquid Crystals, 8 (1969) 285.

PLATELET BIOCHEMISTRY AND FUNCTION -- POSSIBLE USE IN EVALUATING BIOCOMPATIBILITY

C. C. Solomons and E. M. Handrich

Department of Pediatrics
University of Colorado
Denver, Colorado 80220

The thrombogenicity of biomaterials in contact with blood is an important consideration in the design of prosthetic devices for implantation (1). This paper describes the development of a sensitive '*in vitro*' method for studying the energy metabolism of cells under various conditions of stress. The technique is presented in detail for blood platelets but can be applied, with modifications, to any population of cells which produces energy in the form of ATP.

Platelets utilize energy in order to maintain themselves, synthesize molecules, transport ions, and conduct other specialized functions. In performing its function of supplying energy to the cell ATP is degraded to ADP and AMP from which ATP can be regenerated. However if a large amount of AMP is formed by an excessive ATP utilization, then some of the AMP is irreverisbly deaminated to IMP and lost to the energy cycle. Further degradation of IMP to hypoxanthine (HYPX) occurs and results in an appreciable loss of 'expensive' purines from the cell. Thus a shift in the cellular profile of these nucleotides is an index of a metabolic disturbance in energy metabolism. The energy required as ATP for platelet aggregation is generated mainly from glycolysis, which is partly controlled by the enzyme phosphofructokinase (PEK). This enzyme is inhibited by ATP and C-AMP, and stimulated by ADP and AMP.

The physiological function of platelets is to adhere to the exposed surfaces of a cut blood vessel thus initiating the formation of a hemostatic plug. As more platelets accumulate, they aggregate to each other, and participate in the formation of a fibrin clot which stops the bleeding. Healing of the cut then takes place and the clot is ultimately lysed. Many substances can act as a trigger to cause platelet aggregation: collagen, ADP and epinephrine are potent stimulators of aggregation, while prostaglandins and salicylates are inhibitory (2). Normally the platelets have large stores of ATP and ADP which are maintained for use in the aggregation process (3). Thus the energy metabolism of the platelets is a clue to their ability to function when triggered. The occurrence of complete platelet aggregation in vitro requires an initial stimulus sufficiently powerful to cause a secondary release of stored ADP. If secondary release does not occur, aggregation will be incomplete and disaggregation may occur. However platelets subjected to a subthreshold stress by a foreign surface or an incipient pathological process may not have aggregated but may have acquired a changed sensitivity to subsequent aggregation by repeated stresses. Thus a knowledge of the more subtle internal metabolic changes which can be induced in platelets by genetic or environmental factors offer a means of characterizing the effect of an implant or platelets derived from the blood of an intended recipient both before and during the period of implantation.

METHODS AND MATERIALS

Preparation and Incubation of Platelet-Rich-Plasma (PRP)

1-2 ml of blood was obtained by peripheral venipuncture using a plastic syringe and a #20 gauge needle Leaving the needle in place, the first syringe was discarded and a 12 ml syringe containing 1.0 ml of citrate anticoagulent (3 parts 0.1M sodium citrate and 2 parts 0.1M citric acid) was connected to the needle. 9 ml of blood was drawn (total volume 10 ml). The sample was mixed by gentle inversion several times after introducing a 1 ml bubble of air. The blood was slowly expelled into a plastic 15 ml centrifuge tube. For platelet function tests of aggregation another 20 ml of citrated blood was drawn and treated according to standard methods (4). After centrifugation at 900 rpm

(1,500 X9) for 20 minutes at room temperature, the supernatant PRP was gentle transferred using a plastic pipette into a 5 ml plastic test tube and an aliquot (0.2 ml) used for an immediate platelet count. The platelet count was usually $4x10^5$ to $1x10^6$ per mm^3, then 1.0 or 1.5 ml of PRP was incubated at 37°C for 20 minutes; 1 µc of U-14-C adenine was then added in 10 ul and the incubation continued for a definite time period, usually 30 minutes.

Absorption of Nucleotides onto Charcoal (5)

After 14-C-adenine incubation the PRP was centrifuged at 3000 rpm (12 cm rotor) for 3 minutes and the supernatant plasma decanted and saved. 1.0 ml of cold 0.6N perchloracetic acid (PCA) was added to the platelet button and 1.0 ml of 1.2N of cold PCA added to 1cc of the decanted plasma. The tubes are shaken for 15 minutes at 0^oC and centrifuged at 3000 rpm for 5 minutes. The supernatant is kept chilled in ice while the PCA is neutralized with a calculated amount of 1N KOH. (final pH 7-8). The precipitate of potassium perchlorate is removed by centrifugation at 3000 rpm for 5 minutes. The supernatant was then shaken at $+4^oC$ for 20 minutes with 0.1 ml of an aqueous suspension of activated charcoal, Norit A (14 g./100 ml). After centrifugation at 3000 rpm for 5 minutes the supernatant was discarded and the charcoal washed with 2 ml of water and recentrifuged. The charcoal pellets obtained from the platelet-poor-plasma (PPP) and the platelets were shaken with 0.1 ml of 15% pyridine in the cold for 30 minutes to extract the absorbed nucleotides which are then chromatographed on two thin-layer ascending systems.

Chromatography

An aliquot, usually 20 ul of the pyridine extract, is placed on a thin strip (0.1" X 1 1/2") on the starting line of a TLC cellulose support: (Eastman 6064 cellulose-without fluorescent indicator). Another 20 ul aliquot is placed on a Brinkman Polygram Cel-300 PEI chromatogram. Both chromatograms are developed with a mixture of tertiary amyl alcohol; formic acid; water; 3:2:1, and allowed to dry in air, preferably overnight. The first chromatogram (Eastman 6064) is cut into 25 strips 1/4" in width. The strips are cut into 3 pieces and immersed in 5 ml of 'Aquasol' (New England Nuclear) in a counting vial. Radioactivities were measured in a liquid scintillation spectrometer. The second chromatogram is developed a second time, in the same direction, with a mixture of isopropyl alcohol; 0.1M boric acid; concentrated ammonium hydroxide 6:3:1. The solvent is allowed to ascend to the top of the 20th strip before drying in air. The strips are counted as described above. Typical results are shown in Fig. 1.

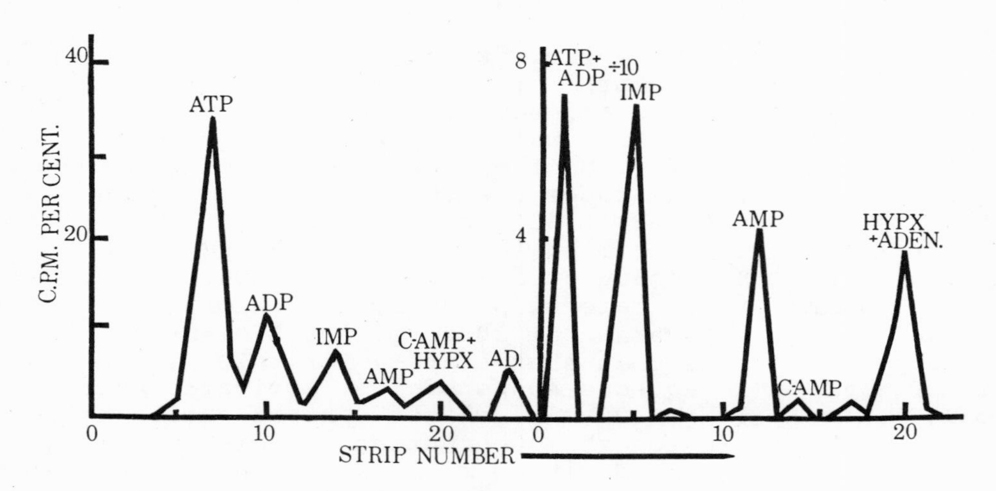

Fig. 1 Chromatographic Profile of Platelet Nucleotides

Calculations

Counting efficiency was uniformly 80-85% including the quenching effect of the Eastman cellulose chromatographic support. Quenching was approximately 5% less for the Brinkman PEI cellulose. Recovery of counts present in the pyridine extract from the chromatograms was 95-100% when quench corrections are applied. For most purposes, the relative amounts of the radioactive nucleotides are calculated as a percentage of the total radioactive nucleotide pool. Thus the radioactivity in cpm of each strip is calculated as a percentage of the total cpm of all of the strips after the activity of the unused C^{14} adenine has been substracted. The percent cpm are plotted against strip number and the percentages under each peak are summed.

CONTROL EXPERIMENTS

Experiments were carried out to determine the sensitivity of the nucleotide profile to changes in the protocol

(1) Platelets

Temperature and time of incubation. Increasing the incubation temperature from 37°C to 41°C decreased platelet ATP by about 15% with increases in ADP (11%) IMP (29%) and AMP(80%). Prolongation of the incubation time from 30-60 minutes had no significant effect on the normal platelet nucleotide profile.

Storage of nucleotides on charcoal. It is often convenient to store the charcoal pellets overnight or longer before eluting with pyridine for chromatography. Investigation of the effects of storage at -20°C showed no significant losses after 8 days. A 15% decrease in ATP and a 37% increase in ADP was seen after 31 days. AMP increased by 90%, IMP and HYPX remained essentially unchanged during this period. In contrast, storage at room temperature for 17 hours resulted in a significant loss of ATP (19%), increases in ADP (80%), AMP (120%) and HYPX (12%). At 30 hours these changes were 53% for ATP, 121% for ADP, 560% for AMP, and 21% for HYPX.

Washing of platelets after incubation. 2 washes with ice cold tris buffer pH 7.4 containing NaCl 0.14M, EDTA 1.5 mM produced a 60% reduction in HYPX alone. Washing at room temperature caused a reduction in ATP (16%) and an increase in IMP (378%) and HYPX (194%). Treatment with glutaraldehyde: This compound immobilizes platelets and is used as a fixative for morphological studies of cells. (6) ATP metabolism is grossly distorted as seen by low values of ATP (28%), HYPX (0%), and high values of AMP (23%) and IMP (18%) and ADP (30%).

(2) Platelet-poor plasma (PPP)

The platelet poor plasma contained about 5% of the platelets still in suspension. 14-C-adenine added to

PPP was not detectably metabolized to form nucleotides. The PPP removed after incubation of the platelets with 14-C-adenine contained radioactive HYPX and unused adenine but no other nucleotides. As the time of incubation was increased to 60 minutes unused adenine in the PRP decreased and HYPX increased.

PATIENT RESULTS

Values are reported for platelets obtained from normal adults and children. See Table I.

TABLE I

Radioactivity of Each Nucleotide as a Percentage of the Total Radioactive Pool of Normal Individuals

	ATP	ADP	IMP	AMP	HYPX	n	In Vitro Platelet Aggregation
Normal Children	61	24	1.5	6.9	4.6	14	Normal
	±9.7	±5.7	±1.0	±3.9	±1.7		
Normal Adults	68	23	2.4	2.5	3.8	11	Normal
	±7.6	±5.3	±1.0	±1.4	±1.0		

DISCUSSION

The method is relatively simple, requires only 5-10 ml of blood and can easily be carried out on patients with a variety of metabolic disorders.

From the control studies it is seen that care must be exercised in handling the platelets expecially with regard to washing with buffer (7) and the temperature during storage of the charcoal pellets. Poisoning of the platelets with glutaraldehyde also leads to appreciable reduction in ATP. Thus washing is usually excluded and storage when necessary is done at -20°C for not more than 6 days. The two systems of chromatography are complementary, and provide corss checks for C-AMP and IMP. C-AMP is separated on the PEI and the cellulose system clearly separates adenine from the other nucleotides. Since HYPX is the only radioactive purine found in the PPP the method offers a means of studying the release of this end product of ATP metabolism.

It should also be possible to adopt the method to measure nucleotide changes affecting the survival of platelets exposed in vitro and in vivo to foreign surfaces such as prosthetic heart valves and other implantable devices. In this way, any genetic peculiarities of the recipient of the implant can be taken into account.

Correlations between the HYPX: ATP ratio and platelet survival measured by tagging with chromium have been obtained in patients with coronary artery disease (8) and these results will be reported at a later date.

SUMMARY AND CONCLUSIONS

1. A method is described in detail for obtaining nucleotide profiles using 1-2 ml amounts of platelet-rich-plasma.

2. The results provide information which was correlated with platelet function in normal individuals, and can be used to study a wide variety of metabolic disorders.

3. The study of the energy metabolism of platelets may become a useful tool in helping to determine the biocompatability of synthetic prostheses using the platelets of the recipient of the implant.

ACKNOWLEDGEMENTS

Supported by Shriners Hospital, Chicago, Illinois, the Easter Seal Research Foundation, N-7223, The Braitmayer Foundation, The Osteogenesis Imperfecta Foundation, NIH Grants NCI-CA 12247, RR-69.

Explanatory note arising from discussion: The ability of this test to quantitate fine differences in metabolism due to subclinical stresses may be used to advantage to increase biocompatability on an individual basis. The properties of a given biomaterial can be modified to match the patient's individual biochemical pattern of platelet function due to his unique genetic tests, dietary habits, medications, and the presence of disease.

REFERENCES

1. Bruck, Stephen D., Macromolecular Aspects of Biocompatible Materials -- A Review; J. Biomed. Mater. Res. Vol. 6, 173-183, 1972.
2. Weiss, Harvey J., Editor; Platelets and Their Role in Hemostasis; New York Academy of Sciences, Vol. 201, 1972.
3. Holmsen, Holm and Day, H. James; Adenine Nucleotides and Platelet Function; Ser. Haemat. Vol. IV, 28-58, 1971.
4. Mull, Marilyn M. and Hathaway, William E.; Altered Platelet Function in Newborns; Pediat. Res. 4, 229-237, 1970.
5. Murakami, M. and Odake, K.; Adenine Nucleotide Metabolism of Human Platelets; Thrombosis et Diathesis Haemorrhagica, Vol. 25, 223-233, 1971.
6. White, James G.; Fine Structural Alterations Induced in Platelets by Adenosine Diphosphate; Blood, Vol. 31, #5, 604-622, May 1968.
7. Waller, H. D., Lohr, G. W., Grignani, F. and Gross R.; Uber den Energiestoffwechsel normaler menschlicher Thrombozyten; Thrombosis et Diathesis Haemorrhagica, Vol. 3, 520-547, 1959.
8. Solomons, Clive C., Steele, Peter; Handrich, E. M.; Unpublished Data.

THE EFFECT OF MICELLES ON ENZYMATIC REACTIONS

Fred R. Bernath

Department of Chemical and Biochemical Engineering
Rutgers University
New Brunswick, New Jersey 08903

INTRODUCTION

It has long been recognized that certain similarities exist in the mechanisms of micelle formation by low molecular weight surface active compounds and the formation of a protein's tertiary structure. These similarities arise from the fact that both surfactants and proteins are amphipathic in nature, i.e. each contains both hydrophilic and hydrophobic moieties. It is clear today that this common characteristic is to a large degree responsible for some of the unique properties exhibited by enzymes, antibodies, lipids, bile salts and other amphipathic substances.

Although the exact structure of micelles and of globular proteins depends on the balance of all types of noncovalent interactions, the most important factor in fixing the general structural characteristics of amphipathic compounds in aqueous solution is probably hydrophobic interactions (1). Studies of this phenomenon have provided a common link between the work of colloid and protein chemists. Due largely to their studies of ionic micelles, colloid chemists were probably the first to appreciate the importance of these interactions. In 1936, for example, Hartley (2) provided a description of micelle formation that is strikingly similar to today's discussion of hydrophobic "forces."

The importance of hydrophobic interactions in the development of the globular structure of proteins was first suggested by Bresler and Talmud (3) in 1944. In 1951 Linderstrom-Lang (4) suggested the terms primary, secondary and tertiary to describe the hierarchy of protein structure and stated that tertiary structure is significantly influenced by the same forces that unite carbon chains in micelles. In 1959 Kauzmann (5) provided strong support for the concept of hydrophobic interactions as the most important factor responsible for globular protein structure. Since that time many other reports have substantiated the concept (1,6).

In 1964 Fisher (7) demonstrated that a protein's three-dimensional structure could be correlated by considering only its relative polarity, i.e. ratio of polar to nonpolar residues. This result provided additional evidence for the extremely close analogy that exists between the structure and properties of micelles and those of globular proteins in aqueous solution. This analogy has been invaluable to the study of protein denaturation and protein structure since the wealth of results that exists for the less complicated micelle systems has been directly applicable to proteins.

The strength of the concept that micelles are excellent models for enzymes has been significantly increased by the discovery that certain miscelles exhibit a catalytic activity which has several of the characteristics of enzyme catalysis (1). Of course, we do not expect the analogy between micelle and enzyme activity to be as strong as that between micelle and enzyme structure. Enzyme activity does not depend only on its amphipathic nature and resulting tertiary structure but also on a specific genetically coded primary amino acid sequence (actually responsible for the enzyme's overall conformation) and a stabilizing secondary structure. However, micelles may still serve as valuable models in studying enzyme activity just as they have aided studies of protein structure. Micelles share some of an enzyme's catalytic properties, particularly those associated with microenvironmental effects in which ion concentration, pH, substrate concentration, and dielectric constant at the surface of the catalyst differ from conditions in the bulk solution. Investigation of the catalytic performance of micelles may, therefore, provide an evaluation of the magnitude and relative importance of these common mechanisms as well as insights into the modes of interaction between substrate and catalyst. To date only a few examples of this type of comparative analysis have been presented in the literature. In the future it may prove valuable to apply specifically designed micelle systems as partial catalytic models of specific enzymes.

As mentioned earlier, hydrophobic interactions have provided a common link of interest between those studying the structure and properties of micelles and those investigating the same characteristics of proteins. Although the theories of micelle formation have proved helpful to protein studies, developments in each area have, in general, proceeded concurrently and independently. There has been one area, however, in which a significant overlap of effort has occurred. This has been in the investigations of the effects of synthetic detergents on protein denaturation (8). These studies have provided descriptions of the characteristics of detergent - enzyme interactions and have also given insights into the nature of protein - protein and protein - substrate interactions.

Early studies seem to indicate that the primary mode of detergent - enzyme complexing is electrostatic in nature. (8) It is obvious today that electrostatic forces are only one of a number of noncovalent interactions and that it is the overall balance of these forces including hydrogen bonds, electrostatic forces, van der Waals - London dispersion forces and hydrophobic "forces" that determines not only the structure of micelles and enzymes but also the interactions that occur between them. Some have suggested that, in addition to its importance in determining structure, hydrophobic effects are also the primary driving force involved in micelle - protein, protein - protein, and protein - substrate complexing. In the terminology of Jencks (1), the pendulum of opinion at one point swung sharply away from the hydrogen bond in the direction of the hydrophobic "bond" as the primary driving force for intermolecular interactions. Today, however, it is "beginning to swing back to a more central position, although it has by no means come to rest." Perhaps the discussion of "primary" forces is purely academic since enzyme activity and specificity probably depend on a delicate balance of all noncovalent forces, the relative importance of individual components varying for each particular enzyme system.

Early studies also indicated that detergent - enzyme interactions usually result in enzyme denaturation and inactivation. It is obvious today that this is not necessarily a general phenomenon. Detergent micelles bind to enzymes by noncovalent forces of the same type that are responsible for enzyme - substrate, enzyme - inhibitor, and enzyme - activator complexes. Inactivation is probably caused by a blockage of the active site, a conformational change and/or a change in the microenvironment of the enzyme. There is no reason to expect that the latter two occurrences will always lead to inactivation rather than activation. An enzyme's "native" conformation in dilute aqueous solution is not necessarily its natural conformation nor its most active conformation. Most enzymes probably perform _in vivo_ in gel-like surroundings, in solid state assemblages or while adsorbed at various interfaces (9). The enzyme's condition in dilute aqueous solution, then, may not necessarily be its most active, and conformational or environmental changes may enhance activity as well as destroy it. This analysis has been strengthened by a number of reports of activating or stabilizing interactions between surfactant micelles and enzymes (10, 11, 12).

As mentioned above, most enzymes probably do not perform in dilute aqueous solution _in vivo_. It is also a distinct possibility that the natural condition for an enzyme is not its free form but one in which it is bound to a carrier _in vivo_. One example of a natural carrier may be collagen, the major extracellular structural protein and the most abundant of all proteins in the higher vertebrates. This idea is supported by the recent discovery that reconstituted collagen is an excellent enzyme carrier _in vitro_, binding enzymes by the same cooperative noncovalent interactions that have been previously discussed (13, 14).

Collagen is probably only one of a number of other potential _in vivo_ enzyme carriers. Recent reports suggest that micelles such as those of phospholipids or bile salts may also serve this function. It has been reported that most multienzyme systems which exist in cells require phospholipids for their structure and activity (15) Some have suggested that phospholipids interact with enzymes in the form of micelles (10) which cause activation by creating a conformational change in the complexed enzyme (16) or by immobilizing the enzyme with its active center oriented in some optimum manner (17). It is significant to note that a particular enzyme often exhibits little specificity for the phospholipid. For example, Green et al. (10) have shown that an inactive, almost lipid-free preparation of cytochrome oxidase can be activated either by the addition of phospholipids or the nonionic surfactant Tween 80 (polyoxyethylene sorbitan monoolaete). Other examples of micelle bound enzymes are given by Tettamanti (11) and Takeda and Hizukuri (12) who demonstrate an activating effect of Triton X-100 micelles (polyoxyethylene octyl phenol) on rabbit brain neuramidase and sweet potato -amylase. These studies indicate that surfactant micelles may be useful as carriers for enzymes *in vitro* in order to simulate the natural surroundings of an enzyme or to enhance or stabilize its activity.

OBJECTIVE AND MOTIVATION

For a number of years detergent micelles have served as useful models for the investigation of enzyme structure and activity. Recent experiments have suggested that studies involving micelles may provide still another extremely valuable contribution to the theories of enzyme activity. From the above discussion it appears quite possible that micelles can bind enzymes *in vivo* and *in vitro* and thereby impart an activity or stability enhancement to the enzyme. In view of the general types of noncovalent forces that are responsible, this effect may be a general one. The objective of the work described in this paper is to initiate a systematic research effort to explore this phenomenon.

The objective of the initial study described here is to characterize the effect of nonionic micelles on the lysozyme catalyzed reaction. Lysozyme was chosen as a model enzyme because it is probably one of the best characterized of all enzymes (18). An additional motivation is the recent work by Ossermann which suggests that lysozyme may play a role in the body's general mechanism to resist cancer and other diseases (19). The substrate used in this study was a suspension of dried *Micrococcus lysodeikticus*. A nonionic detergent, Tween 20 (polyoxyethylene sorbitan monolaurate), was used because lysozyme is extremely sensitive to ionic strength (20 and is generally inactivated by ionic detergents (21).

MATERIALS AND METHODS

Hen egg white lysozyme and dried M. lysodeikticus cells were obtained from Worthington Biochemical Corporation of Freehold, N.J. The lysozyme was salt free, twice recrystallized, and had a specific activity of 10,015 units/mg. A unit is defined as the amount of enzyme that causes a decrease in absorbance at 450 nm of 0.001 per minute when 0.1 ml of 0.05 mg/ml lysozyme is added to 2.9 ml of a 300 mg/l cell suspension at 25 C, pH 7.0 and 0.1 M. phosphate buffer. Tween 20 was obtained from J. T. Baker Chemical Company, Phillipsburg, N.J. The surfactant had a mean molecular weight of 1385 and a mean ethylene oxide content of 23.5.

The reaction was followed by observing the rate of clearing of a turbid suspension of dried cells at 450 nm and the above conditions. Measured optical densities were converted to substrate mass concentrations through the use of a calibration chart. A well stirred 600 ml reactor was linked to a sampling system consisting of a peristaltic pump, a colorimeter fitted with a debubbler and a recorder with a special rapid chart drive. The analytical system enabled close monitoring over the first 15 - 100 seconds during which the volume change in the reactor was negligible. Enzyme concentration was varied from 1.67 - 100 mg/l, substrate concentration from 100 - 1000 mg/l, and surfactant concentration from 0 - 5 g/l.

Ultrafiltration studies were conducted using an Amicon model 402 cell fitted with a PM-30 or XM-100 membrane. Ultraviolet data were obtained on a Beckman DB-G grating spectrophotometer. Fluorescence data were obtained utilizing an Aminco - Bowman spectrophotofluorometer with exciting wave length set at 285 nm and emission readings taken at 345 nm. The surface tension isotherm for Tween 20 was obtained with a Rosano Surface Tensiometer employing the Wilhelmy plate method.

RESULTS AND DISCUSSION

It was found that the time course of the reaction could best be described by second order kinetics with respect to substrate. Figure 1 shows second order plots of data for an enzyme concentration of 50 mg/l

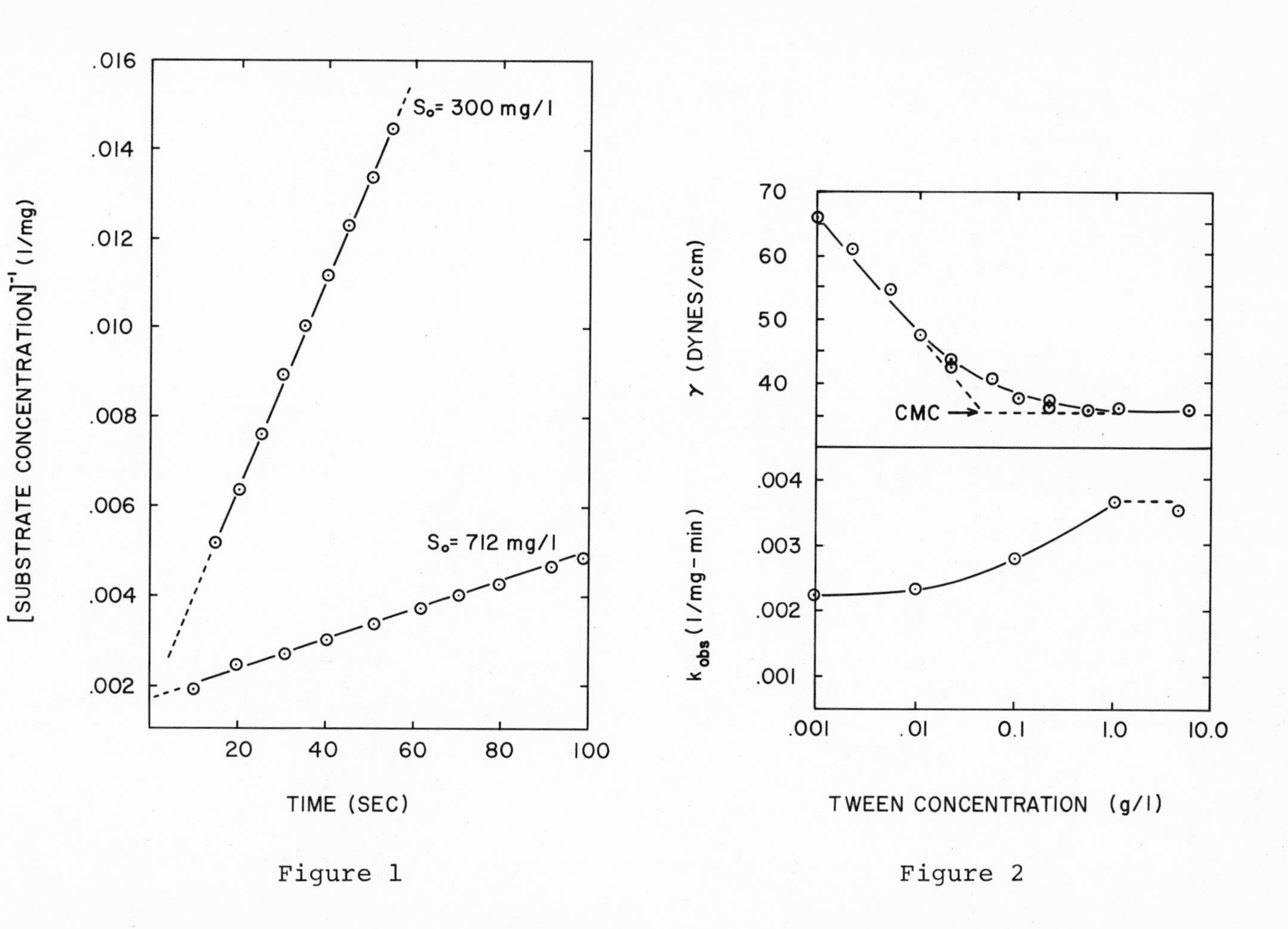

Figure 1

Figure 2

and two different substrate concentrations. Observed rate constants were calculated from the slopes of the straight line portions of these plots and were used to compare enzyme activity under various conditions.

Polyoxyethylene sorbitan monolaurate (Tween 20) is a nonionic surfactant that consists of a sorbitan backbone, three polyoxyethylene chains containing approximately 20 ethylene oxide groups and a laurate (12 carbon fatty acid) chain. Figure 2a shows that its critical micelle concentration occurs in the region 0.05 - 0.1 g/l. Above the cmc the surfactant forms micelles that are probably rod-shaped and contain 20 - 100 monomers (22). The molecular weights of these micelles may be on the order of 100,000 as compared to 14,500 for lysozyme.

Figure 2b shows the effect of the nonionic surfactant on the lysozyme reaction for the case where substrate concentration is 750 mg/l and enzyme concentration is 50 mg/l. Below the cmc an increase in surfactant concentration has little effect on the observed second order rate constant. As the micelle region is approached, however, the rate constant begins to increase with surfactant concentration until well into the micelle region where it levels off. A further increase in surfactant concentration causes the rate constant to decrease somewhat. For the given conditions the maximum effect is a rate increase of 70%. Other results indicate that the rate enhancing effect of Tween 20 is less pronounced (50%) at a substrate concentration of 300 mg/l and that it increases with enzyme concentration for lysozyme levels between 10 and 50 mg/l. No effect is observed below an enzyme concentration of 10 mg/l. The optimum surfactant concentration also appears to vary with enzyme concentration. For a lysozyme concentration of 50 mg/l the optimum Tween 20 concentration is 1 g/l while at 100 mg/l the maximum rate increase is observed at a surfactant concentration of 4 g/l.

Figure 2b appears to discount the importance of individual surfactant molecules since the effect is only observed above the cmc. It also appears doubtful that the rate enhancement is caused by a surfactant micelle - substrate interaction. If such an interaction were taking place, we would expect to observe a rate effect even at low enzyme concentration. We also would not expect a variation in the optimum Tween 20 concentration with enzyme concentration. Additional experiments have shown that Tween 20 has no effect on the rate of autolysis of M. lysodeikticus in the absence of enzyme and that fluorescence emissions for surfactant and substrate are additive when the two are combined in solution.

The results cited thus far appear to strongly suggest that a micelle - enzyme complex is responsible for the rate enhancement effect. The nature of the complex and its mechanism of action are still not obvious, however. Tween 20 micelles probably bind lysozyme molecules by the cooperative noncovalent forces that were discussed earlier. In this instance both hydrogen bonds and hydrophobic interactions are probably important. Since polyoxyethylene surfactants are known to have a high affinity for phenol hydroxyls (23), binding might take place between the ethylene oxide groups of the surfactant and the tyrosine residues of the enzyme which are located away from the active site. In any case, it is reasonably certain that the enzyme is bound at or near the surface of the micelle since water must be available for the hydrolysis of the cell wall to occur.

Micelles could conceivably cause the observed rate effect in one of two ways: by creating an activating conformational change in the enzyme molecule or by binding the enzyme in such a way as to increase the number of productive collisions between enzyme and substrate. If the former mechanism were operating, we would expect to observe a decrease in the activation energy of the reaction. The latter mechanism is an orientation or entropic effect. If it were operating, we would expect to see an increase in the pre-exponential factor but no change in the energy of activation. Figure 3 demonstrates that the rate increase is probably caused by an orientation effect since the presence of Tween 20 micelles does not appear to significantly alter the energy of activation of the reaction.

Figure 4 provides additional evidence for the existence of a micelle - enzyme complex. At 25 C and with surfactant concentrations in the micelle region there appears to be two linear segments in reciprocal substrate - time plots. The first segment coincides with data collected in the absence of surfactant and the second segment shows a higher

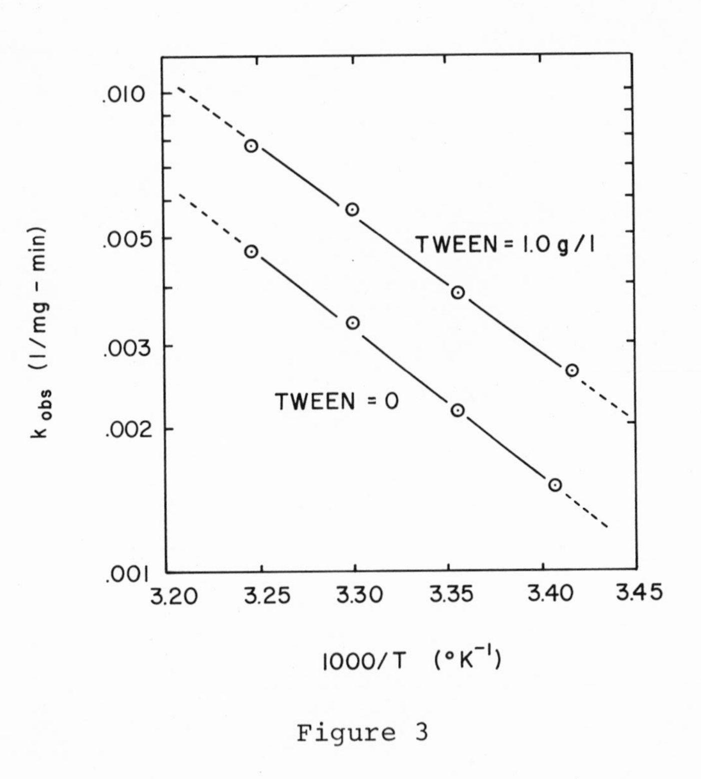

Figure 3

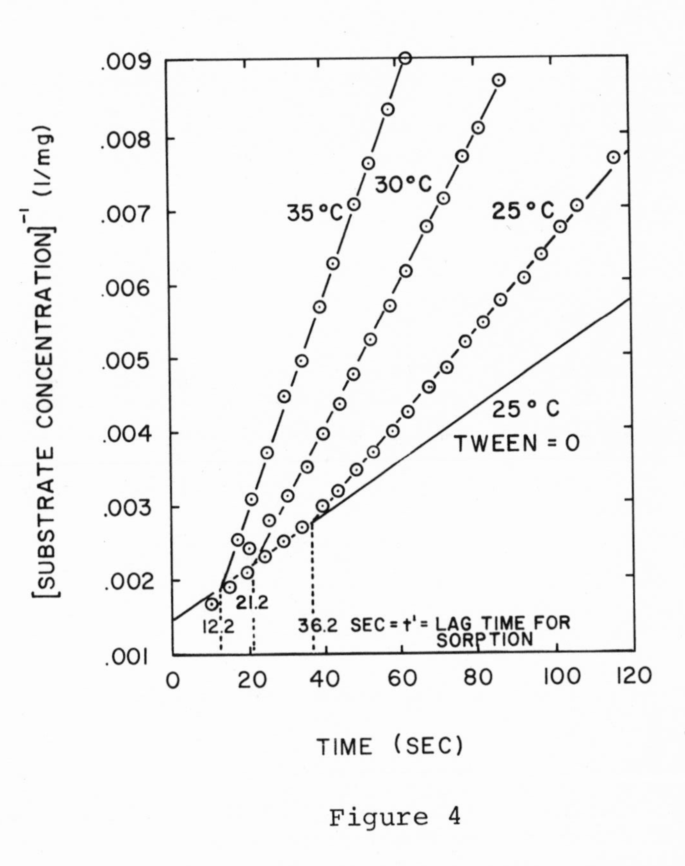

Figure 4

observed rate constant. This kind of behavior would be expected if sorption of the enzyme by micelles proceeded at a rate of the same order of magnitude or less than the initial rate of reaction. In other words, this behavior indicates that there is a lag period during which the enzyme is being sorbed by the micelles. At higher temperatures the rate of sorption increases and the lag period decreases as is expected. If we plot the reciprocal of the lag time versus reciporcal temperature, we get a fairly good straight line, the slope of which is an estimate of the energy of activation for sorption. The value for this system is about 19 kcal/mole which is not unreasonable for a system involving a number of cooperative noncovalent bonds per molecule.

Thus far the existence of an active micelle - lysozme complex has been postulated largely on the basis of kinetic studies. A number of physical studies add further corroboration. Ultrafiltration experiments, for example, demonstrate that a 1.0 g/l solution of Tween 20 is concentrated by an Amicon PM-30 membrane (MW cutoff = 30,000) and passed by an XM-100 membrane (MW = 100,000). This indicates the presence of micelles with a molecular weight somewhere in the range of 30,000 - 100,000. The effect of these micelles on the retention of lysozyme in ultrafiltration through a PM-30 membrane is shown in Table 1. Lysozyme molecules from a pure enzyme solution pass freely through the membrane. In the presence of Tween 20 at micelle concentrations, however, the amount of enzyme retained in the cell is increased by 30 - 40%. These results provide an indication that complexation between micelle and enzyme is taking place.

Fluorescence and UV absorbance data also suggest an interaction between lysozyme and Tween 20. Fluorescence levels, for example, are not additive when lysozyme and Tween are combined in solution. For a particular set of conditions (exciting wavelength = 285 nm, emission wavelength = 345 nm) solutions of pure lysozyme at 50 mg/l gave a relative scale reading of 42.0, pure Tween at 1.0 g/l gave a reading of 4.5 and the combined solution gave a reading of 41.5. For lysozyme solutions of 100 mg/l and Tween 20 at 2.0 g/l the relative readings were 75.0, 7.8 and 74.0. If no interaction were present, we would expect to observe additivity in these readings. UV absorbance readings at 280 nm for the

TABLE I

Effect of Surfactant on Retention of Enzyme in Ultrafiltration

Solutions	Initial Volume	Volume of Filtrate	Lysozyme Activity of Filtrate
Lysozyme = 50 mg/l	150 ml	75 ml	50.5 mg/l
Lysozyme = 100 mg/l	150	73	95.5
Lysozyme = 50 mg/l and Tween= 1.0 g/l	150	74	31.5
Lysozyme = 50 mg/l and Tween= 2.0 g/l	150	74	34.0
Lysozyme = 100 mg/l and Tween= 1.0 g/l	150	75	62.0
Lysozyme = 100 mg/l and Tween= 2.0 g/l	150	70	56.5

An Amicon PM-30 membrane was used. Pressure = 10 psig.

T = 25°C.

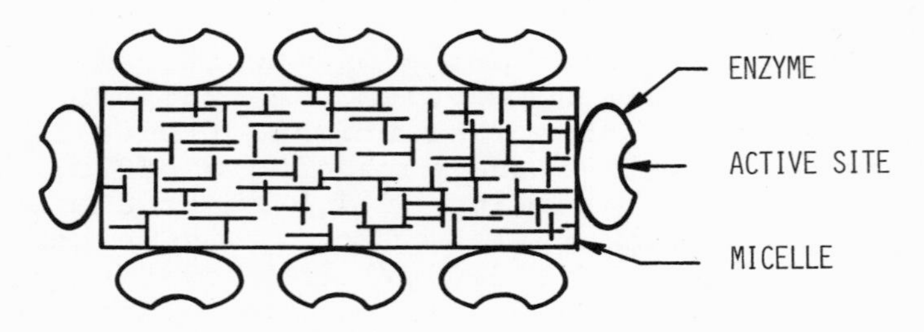

Figure 5

same solutions were additive immediately after mixing. Upon standing at room temperature, however, the absorbance of the combined solution increased faster than that of pure lysozyme. Since the combined solutions retained the same or slightly greater activity than the pure enzyme solutions, the absorbance change is not due solely to denaturation. The order of time for these changes is much longer than that for reaction (order of days as compared to minutes), but the data nevertheless suggest some type of unique interaction between surfactant and enzyme.

Finally, experiments have shown that Tween 20 micelles do not affect the activity of lysozyme when the enzyme is bound to a collegen membrane. Since lysozyme is already held with a specific conformation and orientation on the carrier, this result would be expected if it were in fact the micelle-enzyme complex that was responsible for the observed rate enhancement.

CONCLUSIONS

The nonionic surfactant polyoxyethylene sorbitan monolaurate when present at concentrations above the cmc can cause an increase of up to 70% in the rate of lysozyme action on a suspension of dried cells of *Micrococcus lysodeikticus*. This rate increase is due to the formation of a micelle-enzyme complex in which the enzyme is sorbed in a favorable orientation with respect to substrate particles, thereby increasing the probability of productive collisions. Figure 5 provides a possible description of the micelle-enzyme complex. The combined results of a number of kinetic and physical studies provide convincing evidence for this conclusion.

The result of this work provides encouragement for the further study of micelle-enzyme interactions. For the lysozyme-Tween 20 system more sophisticated experiments, such as spectroscopic, light scattering and electrophoretic studies, would be valuable in characterizing the nature of the surfactant micelles and the micelle-enzyme complex. Other experiments should include studies of the effect of different surfactants on lysozyme, studies of the effect of Tween 20 on other enzymes, and a study of the effect of Tween 20 on lysozyme's action on a soluble model substrate.

Finally, the most important consequence of this study may be the suggestion that immobilized enzymes can be more reactive than the corresponding free enzymes if they have been bound properly. Such a conclusion is not surprising since it has been suggested that immobilized enzymes actually exist in nature. Further studies of specific enzyme-micelle complexes may give some insight into the intrinsic activity of bound enzymes.

The author acknowledges with gratitude the support of the National Science Foundation, Grants-in-Aid GK-10788, 14075 and 29228, and the contributions of his Doctoral Thesis advisor Dr. Wolf R. Vieth.

REFERENCES

1. Jenks, W. P., Catalysis in Chemistry and Enzymology, Chap. 6-8, McGraw-Hill, N. Y. (1969).
2. Hartley, G. S., Aqueous Solutions of Paraffin Chain Salts, p. 45, Hermann et Cie. (1936).
3. Bresler, S. E. and D. L. Talmud, Dokl. Akad. Nauk SSSR 43, 326 and 367 (1944).
4. Linderstrom-Lang K., Selected Papers, p. 410, Academic Press, N. Y. (1962).
5. Kauzmann, W., Advan. Protein Chem. 14, 1 (1959).
6. Vol'kenshtein, M. V., Enzyme Physics, Plenum Press, N. Y. (1969).
7. Fisher, H., Proc. Natl. Acad. Sci. U.S. 51, 1285 (1964).
8. Putnam, F. W., Adv. Protein Chem. 4, 80 (1948).
9. James, L. K. and L. G. Augenstein, Adv. Enzymology 28, 1 (1966).
10. Green, D. E. et al., Arch. Biochem. Biophys. 119, 312 (1967).
11. Tettamanti, G. et al., Enzymologia 39, 65 (1970).
12. Takeda, Y. and S. Hizukuri, Biochim. Biophys. Acta 268, 175 (1972).
13. Vieth, W. R. et al., Trans. N.Y. Acad. Sci. 34, 454 (1972).
14. Venkatasurbamanian, K. and W. R. Vieth, J. Ferm. Tech. 50, 9 (1972).
15. Dawson, R. M. C. and P. J. Quinn, Adv. Expr. Med. and Biol. 14, 1 (1971).
16. Tzagoloff, A. et al., Arch. Biochem. Biophys. 119, 312 (1967).
17. Katchalski, E. et al., Adv. Enzymology 34, 445 (1971).

18. Chipman, D. M. and N. Sharon, Science 165, 454 (1969).
19. Osserman, E. F., Nature, in press (1973).
20. Chang, K. Y. and C. W. Carr, Biochim. Biophys. Acta 229, 496 (1971).
21. Smith, G. N. and C. Stocker, Arch. Biochem. Biophys. 21, 383 (1949).
22. Becher, P., "Micelle Formation in Aqueous and Non-aqueous Solutions," Nonionic Surfactants, M. J. Schick, ed., Marcel Dekker, Inc., N. Y. (1967).
23. Elworthy, P. H. et al., Solubilization of Surface-Active Agents, Chapman and Hall LTD, London (1968).

INTERACTIONS BETWEEN POLYMERS AND MICELLAR LIPIDS*

H. E. Marsh, Jr. and G. C. Hsu

Jet Propulsion Laboratory
Pasadena, California

C. J. Wallace

South Dakota School of Mines and Technology
Rapid City, South Dakota

D. H. Blankenhorn

USC School of Medicine
Los Angeles, California

INTRODUCTION

In the conduct of an investigation with quite practical objectives, a new class of polymers was developed that has the capability of absorbing both water and oils. Study of the unusual interactions of these polymers with micellar lipid solutions led to the belief that there may be clues in the data which could provide new insight on lipid-polymer interactions in general. The purpose of this paper is to examine some physical-chemical aspects of the data from this investigation and propose a qualitative model. For

*This paper presents the results of one phase of research carried out at the Jet Propulsion Laboratory, California Institute of Technology, under Contract No. NAS7-100, sponsored by the National Aeronautics and Space Administration.

orientation, a brief review of the practical goals and results follow directly.

A frequently used treatment for hypercholesterolemia operates on the principle of reducing the sterol pool by causing fecal excretion of bile acids. This treatment is effective because bile acids are liver-produced derivatives of cholesterol. The agents used in this method are basic ion-exchange resins, which, when ingested, bind bile acids in the intestine. Cholestyramine, a quaternary ammonium styrene-divinylbenzene copolymer, is the best known of such agents.

With the expectation that the direct removal of cholesterol itself by the same mode of administration might be beneficial, we began research aimed at developing polymers capable of binding lipids in general, including cholesterol. We chose absorption as the binding mechanism rather than ion-exchange. The general character of the polymers envisioned for this purpose was lightly crosslinked networks of amorphous chains that would be compatible with lipids.

Although many lipids have both polar and non-polar, or oleophilic, moieties, the latter was considered at first to be most important. For this reason, the early work was done with oleophilic polymer compositions. Formulating studies produced polymers capable of absorbing more than twenty times their dry weight of such lipids as oleic acid and vegetable triglycerides (1). In those tests, the lipids were neat. The results were entirely different when the tests were made with aqueous micellar solutions of lipids. The same polymers were capable of absorbing only 2% of lipids from a synthetic bile containing 45.5% lipids. Clearly, the polar moieties of lipids had to be considered also. Polymers were prepared according to the same plan, but with part of the oleophilic prepolymer replaced by hydrophilic prepolymer. This change gave the desired result (2). Lipid absorption as high as 50% (based on original polymer weight) was attained with one composition of this new type of polymer from synthetic bile. Subsequent preliminary *in vivo* tests, in which polymer was mixed with the feed of rats, showed a significant increase in fecal neutral steroid-excretion.

MATERIALS

Polymers

Materials. Polymers were prepared by urethane cures in solvent. Mixtures of nominally bifunctional hydroxyl-terminated prepolymers and trihydroxy crosslinking agents were cured with diisocyanates. The prepolymers used were polybutadiene (General Tire and Rubber's Telagen, 2000 molecular weight, and Phillips Petroleum's Butarez, 4000 molecular weight) for oleophilic, polyethylene oxide (1000 molecular weight) for hydrophilic, and polypropylene oxide (2000 molecular weight) for intermediate affinity. Trimethylol propane and 1,2,6-hexane triol were used as crosslinking agents. Two curing agents were used, 2,4-tolylene diisocyanate and dimer acid diisocyanate (600 molecular weight). Ferric acetyl acetonate was used as a catalyst when necessary. Both dioxane and benzene were found to be suitable solvents.

Polymer Preparation. Formulations were generally calculated to contain 0.25 equivalents of crosslinking agent, 0.75 equivalents of prepolymer and from 0.8 to 1.2 equivalent of isocyanate. Solvent-to-polymer ratio was around 3 to 1. Ingredients were weighed, mixed and heated until homogeneous, and then stored covered in an oven at 71°C for several days to approach complete cure. After cure, all of the sol fraction was removed by soxhlet extraction or by repeated immersion in solvent. Sol removal was followed by exposure to vacuum, for solvent removal. Size reduction was accomplished in a small blade grinder. Grinding was tried at various stages after cure. It was found that there were certain advantages to doing some of the grinding early in the sol removal step, with the polymer swollen by solvent.

Micellar Solutions

Two types of isotropic bile were used. A synthetic bile consisted of 54.5% water and 45.5% lipids: sodium cholate/crude egg lecithin/cholesterol 55/44/1. The other bile was prepared by reconstituting dehydrated ox bile with water at varying concentrations and filtering. All bile solutions were clear and isotropic when used.

RESULTS

The minimum requirements for definition of the class of polymers found capable of absorbing lipids from micellar solutions in significant amounts are the presence of both oleophilic and hydrophilic segments and the existence of an amorphous state. The latter condition may not exist in the neat polymer but must at least, when the polymer is dilated with appropriate solvents. The several types of segmented polymers developed by Lyman et al (3,4,5) probably fit in this class. Two more conditions were added for purposes of the present study. They are crosslinking and absence of sol fraction.

Effect of Polymer Composition

Polymers containing varying proportions of oleophilic and hydrophilic chain-material were ground and contacted with bile solutions having various compositions. The ground products were in the form of very irregular particles about 1 mm in size, which was the best size reduction attainable with these rubbery materials in the grinder without resorting to cooling below the glass transition temperature. Near equilibrium absorption was reached with this size of particle in about 144 hours, as indicated by rate measurements discussed below. Near equilibrium absorption results from these tests are illustrated in Figures 1, 2 and 3. In these figures, the abscissa represents chain material weight fractions only. Small molecules, like trimethylol propane and tolylene diisocyanate, were not counted. Neither were end groups, reacted or unreacted. For those formulations containing dimer acid diisocyanate instead of tolylene diisocyanate, the hydrocarbon portion of the molecule was counted as oleophilic chain material.

The data in Figures 1 and 2 demonstrate clearly the need for a mix of both kinds of chains to absorb significant amounts of micellar lipids. To put it another way, adjoining regions (segments, blocks, domains) in the polymer must be compatible, respectively, with the two moieties (polar and non-polar) of the active lipids which hold micellar structures together. Without the existence of this condition, the aqueous micelle represents a more stable environment than the polymer for lipid molecules. Also evident in Figures 1 and 2 is a systematic dependence of lipid absorption capacity

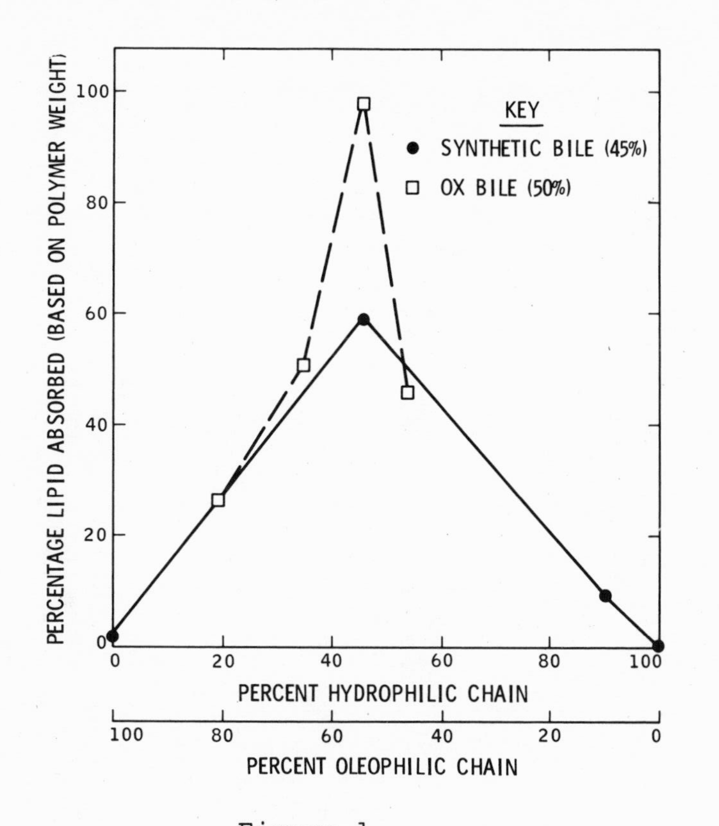

Figure 1
Effect of polymer composition on lipid absorption capability at near equilibrium, 144 hours.

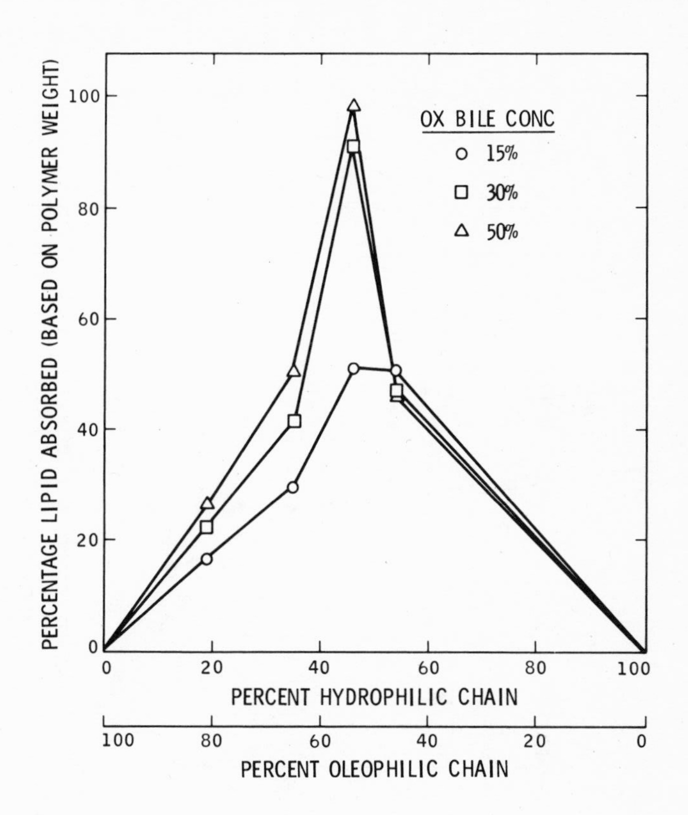

Figure 2
Effect of bile concentration on lipid absorption at near equilibrium, 144 hours

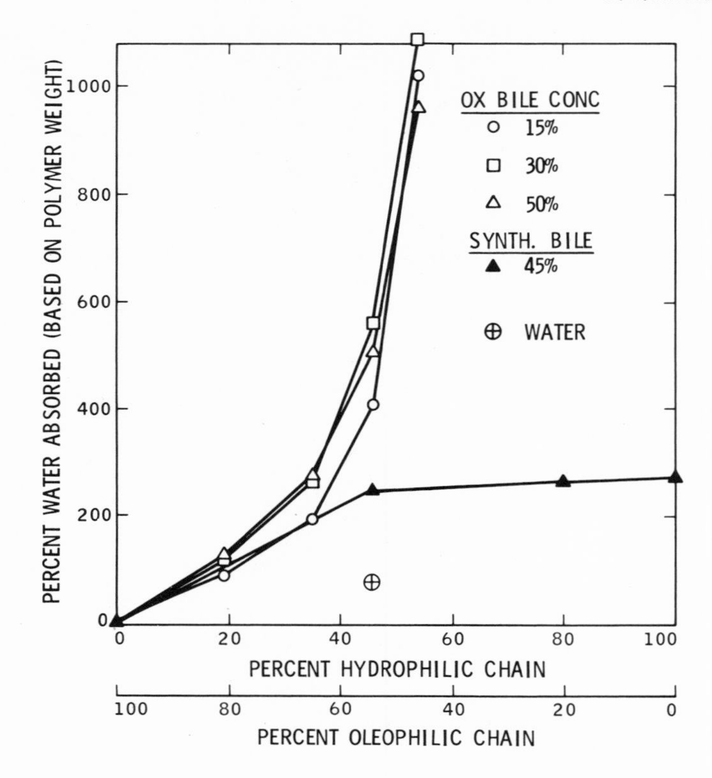

Figure 3. Effect of polymer composition and bile concentration on water absorption at near equilibrium. Notice enhancement of water absorption by the presence of lipids.

on the relative proportions of the two kinds of chain in the polymers. The fact that maximum lipid capacity occurs when the relative weight fractions of oleophilic and hydrophilic chain-materials are about equal will be discussed below in terms of a qualitative model.

Figure 2 shows that, in the range 15% to 50% lipids in ox bile, the lipid concentration effect on polymer absorption of lipids is not very large, except in the case of 54/46 oleophilic/hydrophilic polymer. Here, absorption jumps from near 50% when 15% bile is contacted to near 90% in the case of 30% bile, and then, only an additional 8% when the bile concentration is raised to 50%. Considering the fact that equilibrium was not achieved, but only approached, in these measurements, little weight can be assigned to the absolute values of these differences. On the other hand, a similar difference, shown in Figure 1, between the absorption by the 54/46 oleophilic/hydrophilic polymer from two different types of bile solutions at about the same concentration deserves comment. It is quite possible that the differences in composition of the two biles contribute to the differences in absorption. This observation is reinforced by the difference also in the water absorption from the two biles, illustrated in Figure 3. Although it can only be a speculation, it is worthwhile suggesting that the difference between the state of the bile acids in the two types of bile may be a critical factor. In natural bile, most bile acids are conjugated with taurine or glycine; on the other hand, cholic acid, free from conjugation, was used to prepare the synthetic bile. This one factor accounts for one of the largest differences between the phase relationships of the systems water/lecithin/ sodium cholate and water/lecithin/bile acids (conjugated). (6). In the former, isotropic compositions do not exist above the range 55 to 70% lipids, depending on lecithin concentration; in the latter, the limit is above 90%.

One set of data, shown in Figure 3 was most unexpected. At near equilibrium the 54/46 oleophilic/ hydrophilic polymer in contact with synthetic bile absorbed three times as much water as when it was contacted with pure water. The multiple for the sample polymers in contact with ox bile was six. The hypothesis, that the restriction in water absorption capacity in the case where pure water was used was caused by mechanical restraint of the unswollen oleophilic fraction through interconnecting crosslinks, was not proved in a simple experiment. A sample of the polymer was agitated with a mixture of water and hexane until equilibrium was reached. The total absorption was equivalent to the sum of that for the two solvents alone. No enhancement of water absorption occurred. At this time, we have no explanation for this behavior.

Compatibility Considerations

The legitimate questions arises as to whether the observed phenomena are caused, as proposed, by the presence of adjoining regions in the polymer with specific dissimilar compatibilities, or by the overall compatibility of the polymer resulting from the mixture of two kinds of chain material. This question was investigated by comparing the micellar lipid absorption capabilities of two polymers with nearly equal bulk solubility parameters but representing the two types of structures under consideration. One polymer was based on hydroxyl-terminated polypropylene oxide; it was made the same way as the mixed-chain polymers, including concentration of crosslinking agent and ratio of isocyanate to hydroxyl. The other polymer was one of the mixed-chain polymers whose solubility parameter was estimated by calculation to be nearly the same. Near equality of solubility parameters was confirmed by equilibrium swelling in a series of solvents with known solubility parameters ranging from 7 to 23. A good match is indicated, in Figure 4, by the occurrence of both maxima between 9 and 10. It also appears that the crosslink density of the polypropylene oxide polymer was nearer half that of the 54/46 oleophilic/hydrophilic polymer than equal to it, as planned. If bulk compatability is the critical factor for micellar lipid absorption, the polypropylene oxide polymer would then be expected to absorb more lipids. However, this is not the case. The polypropylene oxide polymer absorbed only 0.98% lipids in 144 hours, while the mixed-chain polymer in this test absorbed 39.4%.

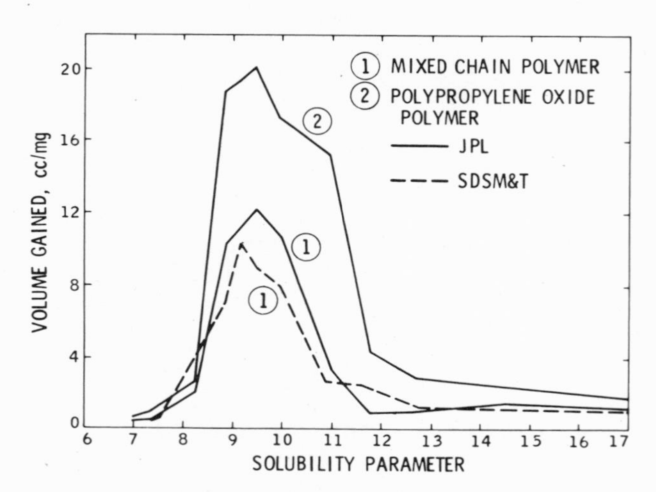

Fig. 4. Determination of solubility parameters of polymers by swelling tests. Samples of polymers were soaked in ten different solvents, each representing a different value of solubility parameter.

Thus, the hypothesis that dissimilar regions with specific compatibilities are required is proved. If this is considered in terms of polymer segments, the question of how large the segments must be arises. As far as this investigation goes, a partial answer can be given. Segments ranging in molecular weight from 600 to 4000 are large enough. The polypropylene oxide chain can be thought of as a series of hydrophilic-like segments, centered around each oxygen molecule, and oleophilic-like segments alternating with them. Obviously, such segments are too small. Critical segments size must lie between these two limits.

Two further comments can be made on these results. For some cases of physical interaction involving molecules and polymers with dissimilar moieties or regions, regional compatibility factors (such as solubility parameters) may be more meaningful than bulk ones. Another way of describing the mixed-chain polymers of this study would be to call them cross-linked nonionic surfactants.

Rates of Absorption

The rates of bile absorption were studied by subjecting weighed samples of the same polymer to contact with samples of the same bile for various lengths of time. (Bile samples were large enough to ensure essentially constant composition.) For each set, polymer samples were separated from bile and weighed as soon as possible after end of contact. Lipid absorption for each set was then determined by removing absorbed water (vacuum drying) and weighing again. The absorption curves in Figures 5 and 6 are typical. The fact that all three lipids were absorbed was verified by thin layer chromatography tests of extracts taken from vacuum dried samples.

Clearly, the process of absorption of bile lipids from micellar solutions by mixed-chain polymers is a very slow one. The data indicated that even in 144 hours, equilibrium had not been reached in 1 mm. particles. With 6 mm. chunks, the process is even slower. The wide difference in absorption rates of the lipid fraction between particles and chunks suggests that diffusion of lipid molecules is the rate controlling step in the process. With the water

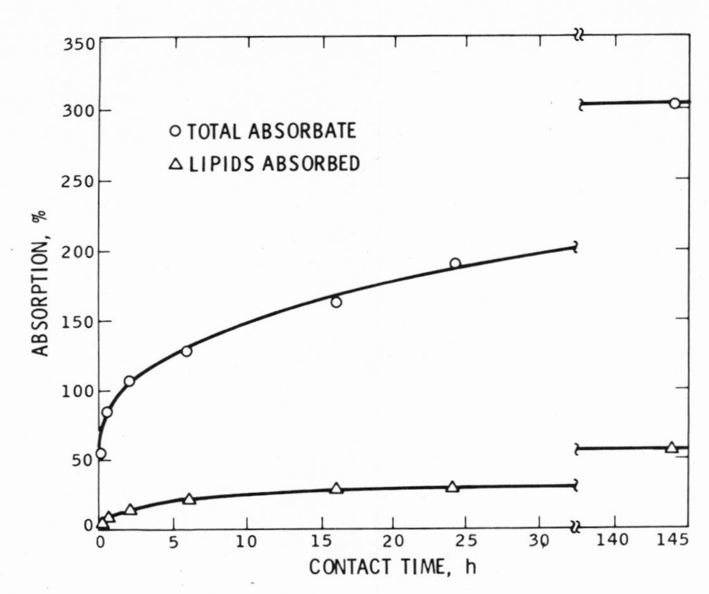

Fig. 5. The rate of absorption of synthetic bile by ∿ 1 mm particles of 54/46 hydrophilic/oleophilic polymer. The total absorption includes water. The lipid curve data were determined by weighing after drying.

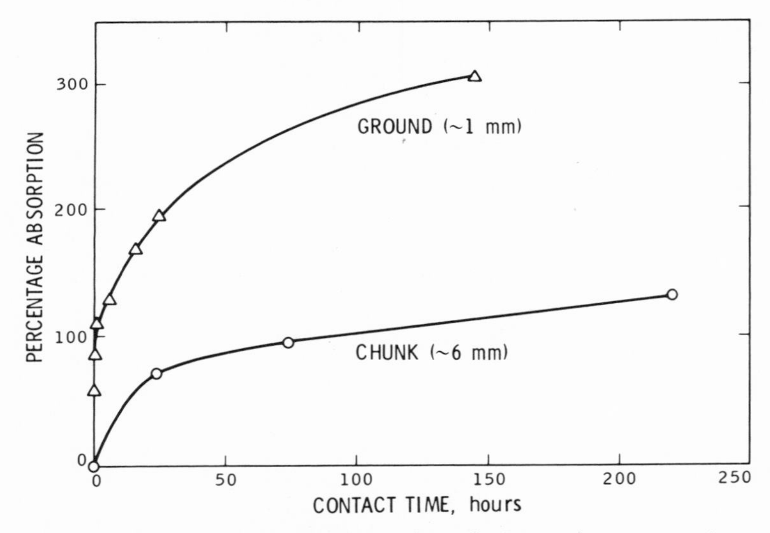

Fig. 6. Effect of polymer particle size on the rate of synthetic bile absorption. Lipid absorption curves (not shown) have the same relative positions as these total absorbate curves.

fraction, it is a different matter. Water concentration in the polymer usually reached its neat equilibrium concentration in from 5 to 10 minutes. From that point on, the unknown mechanism of lipid enhancement of water absorption took over.

Desorption

Of the solvents tests, dioxane was found to be the most effective in removing lipids from dried test polymers. Neither benzene nor water seemed able to remove all of the lipids. This limitation of water is illustrated in Figure 7. In all cases in which water was used for desorption, the residual, unextracted lipid concentration was approximately one-third of the original value. The other important observation in these tests is that the initial rates of water and lipid leaching were considerably higher than initial absorption rates. This suggests that the concentration of these materials in the polymer surface layer was considerably higher than in the interior.

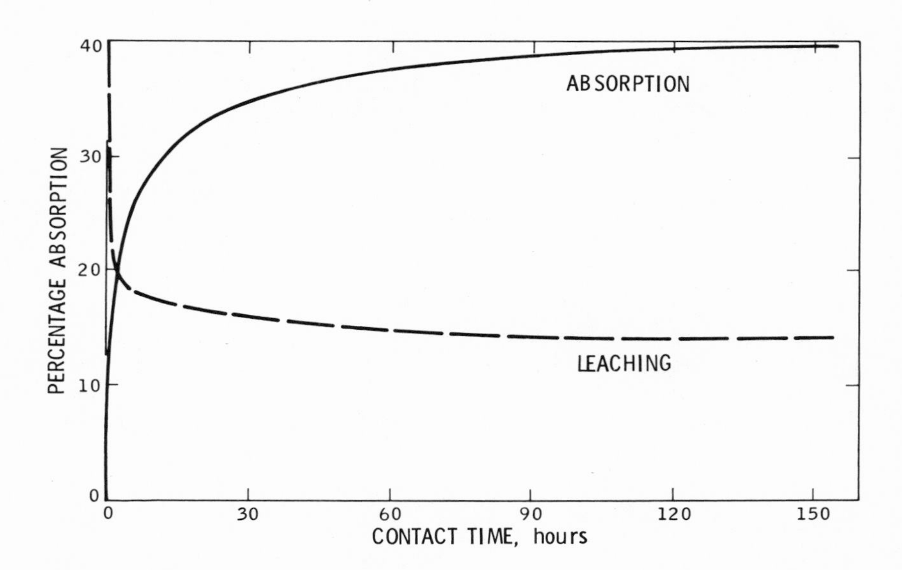

Fig. 7. Comparison of the rate of absorption and leaching, for lipid content only. Solution was 50% ox bile during absorption and was water during leaching.

A QUALITATIVE MODEL

The scope of this investigation was limited, and even within these bounds, the work was preliminary. Nevertheless, the character of the results is of such a nature that it seems appropriate to extend some of the conclusions beyond the present scope to other crosslinked, amorphous polymers and other aqueous micellar solutions. The conclusions of the present study are summarized as follows:

1. Hydrophilic polymers absorb very small quantities of lipids (∿1% in this study) when placed in contact with aqueous micellar solutions of lipids.

2. Oleophilic polymers also absorb small quantities of lipids. (However, in this study oleophilic polymer absorbing capability was about twice that of hydrophilic polymer.)

3. Polymers containing both oleophilic and hydrophilic segments, homogeneously distributed, (called mixed-chain polymers) are capable of absorbing large amounts of lipids. (In this study, close to 100%, based on original polymer weight, was achieved.)

4. Water absorption accompanies lipid absorption when mixed-chain polymers are placed in contact with micellar lipid solutions. (In this study, water content exceeded lipid content.)

5. Lipid absorption capability of mixed-chain polymers depends systematically on the relative fractions of the two kinds of chain. (In this study, maximum capacity occurred at around the 50/50 composition.)

6. There is a minimum segment size for mixed-chain polymers below which significant lipid absorption will not take place. (Results of this study indicate the minimum could be between 30 and 600 in molecular weight units.)

7. The lipid absorption process is very slow. (In this study, equilibrium between polymer, in ∿1 mm. particle size, and bile solution was not reached in 144 hours.)

8. Conversely, the absorption rates for small molecules (water and hydrocarbons) is very rapid. (In the case of the mixed chain polymers of the present study, in contact with bile solutions, water content reaches the level found for neat water contact in 5 to 10 minuts.)

9. The capacity of the polymer for water is enhanced by the presence of lipids. (One composition in this study absorbed from 3 to 6 times as much water from bile solutions as it did from pure water.)

10. It appears that the controlling step in the process of micellar solution absorption is the diffusion of lipid molecules in the polymer.

11. The process of leaching previously absorbed bile lipids from mixed-chain polymers by water follows a pattern which indicates the possibility that rate is controlled by two different mechanisms, one after the other. (In this study, initial lipid leach rate was much faster than initial absorption rate. This was followed by a very slow leach rate, and the water appeared to be unable to extract more than about 2/3 the previously absorbed lipids.)

The following qualitative model attempts to account for as many as possible of the above observations. The model is in two parts. The first part describes the structure of crosslinked, amorphous polymers. The second part describes the process that occurs when such polymers are put in contact with micellar solutions.

Both solvent and heat were found necessary to obtain, and retain during curing, clear mixtures of the two prepolymers and other ingredients. Finished mixed-chain polymers were translucent, indicating some kind of phase separation. Since the curing mixtures were not stirred, it is reasonable to expect the phases to be aggregates of the two kinds of material, thus producing a homogeneous, but random, distribution of oleophilic and hydrophilic domains. The domains are microscopic or smaller. The question is how much smaller. A limiting case, derived from the research of Vollmert and Stutz, (7) is shown in Figure 8. This model is applicable to both mixed-chain and single-chain polymers. In this study, each domain may be a single prepolymer molecule, or it may be an aggregate. Also shown in the figure are the

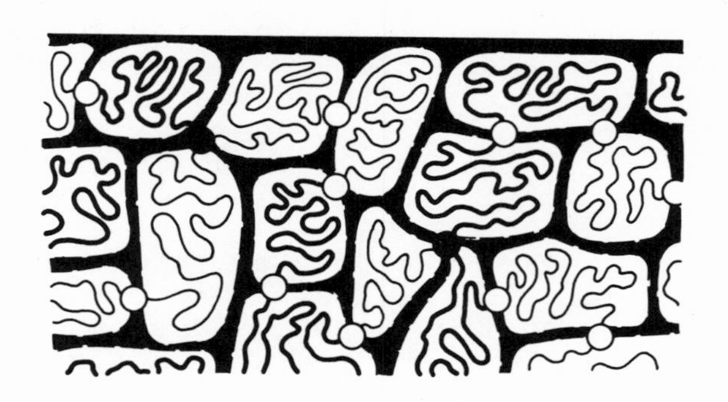

Fig. 8. Possible morphology of crosslinked, mixed-chain polymer. In this model, which is derived from Vollmert and Stutz (Ref. 7), the smallest possible domains are assumed, each one being a single prepolymer chain. Also shown are the chain extending (white circles) and crosslinking (black circles) connections of the network.

chain-extending and crosslinking connections which tie the structure together.

Now, let us consider what happens when a crosslinked, amorphous polymer is placed in contact with an aqueous micellar solution of bile lipids. Before doing this, it is necessary to describe briefly the structure of a micelle because of its importance to the process. A micelle in water is a sphere-shaped aggregate of a small number of molecules, a minimum fraction of which have both hydrophilic and hydrophobic moieties, the remainder being hydrophobic, organized so that the hydrophobic fraction is surrounded by a shell made up of the polar fraction. In bile, the bile acids and lecithin are the polar-nonpolar molecules. By comparison, the polar nature of cholesterol is very low.

Consider first the process when an all hydrophilic polymer is contacted with a micellar solution. Upon immersion, the polymer quickly imbibes water and swells. The polar shell of a micelle that, by chance, contacts the swollen polymer surface finds the water in the polymer and the water in the solution to be nearly alike. After equilibrium is reached, the polymer surface will have a constant overall concentration of fully organized micelles loosely attached to it. In the rinsing operation used in making the measurements of this study, most of the lipids so attached would be removed. This arrangement is illustrated in Fig. 9.

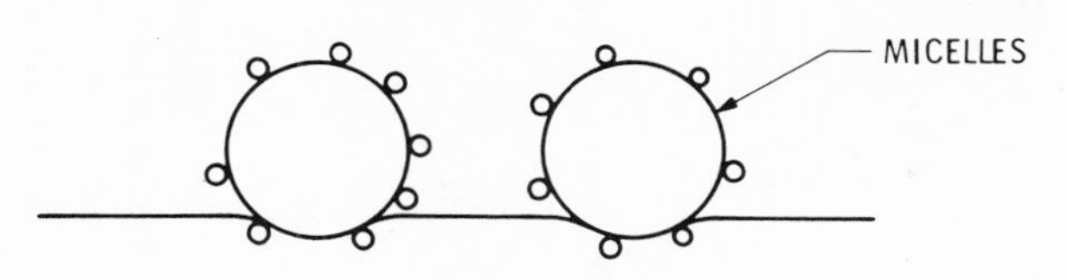

Fig. 9. Representation of the interaction of a micellar lipid solution in contact with a hydrophilic polymer. Undisrupted micelles associate loosely with water-swollen polymer surface. Large circles represent the hydrophobic parts of the lipid molecules aggregated and surrounded by the hydrophilic parts, the small circles.

In the case of an oleophilic polymer, the process is different. No water enters the polymer. When a micelle comes in contact with the polymer surface, one of two things happens. Either the micelle migrates back into the solution away from the surface which appears hostile to the hydrophilic shell, or the micelle disrupts, and the lipids spread out over the surface, because of the tendency of polar lipids to encapsulate hydrophobic material. The ultimate consequence of this process (at equilibrium) is a distribution of polar lipids at the polymer surface in equilibrium with the micelles of the solution. In this state, each polymer particle could be visualized as a very large micelle. This arrangement is illustrated in Figure 10.

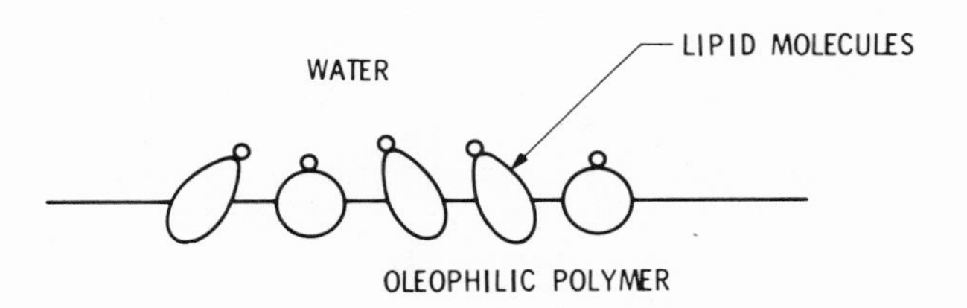

Fig. 10. Representation of the interaction of a micellar lipid solution in contact with an oleophilic polymer. Micelles disrupt, and the individual lipids associate with the polymer surface. Polar moieties (small circles) remain in water phase. Hydrocarbon parts (large circles and ovals) diffuse into surface polymer.

The process which occurs when a mixed-chain polymer contacts a micellar solution starts out as a combination of elements of the two just discussed. First, as illustrated in Figure 11, the polymer swells by the rapid absorption of water into the hydrophilic domains. The chains in the oleophilic domains are forced to move about and extend during this initial swelling because of their interconnection with the hydrophilic ones. In view of the polymer model in Figure 8, it is not at all clear how water finds its way to all hydrophilic domains. Certainly, the ease with which this occurs depends on the relative concentrations of the two kinds of domains. Micelles that come in contact with the polymer at the surfaces of hydrophilic domains will act as described previously.

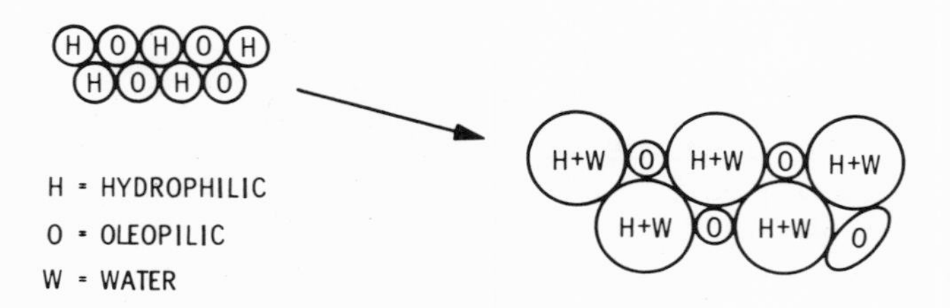

Fig. 11. Representation of the first step in the process which occurs when a micellar lipid solution is contacted with a mixed-chain polymer. As in Figure 9, water is rapidly absorbed by the hydrophilic domains. Notice the possibility of strain imposed on some oleophilic domains by this step of the process.

Also, as previously described, micelles that contact oleophilic domain surfaces become disrupted, and the lipids distribute themselves individually at the surface. In mixed-chain polymers, the process doesn't stop here. The key to absorption is presence of interfaces between hydrophilic and oleophilic domains. As illustrated in Figure 12, these interfaces provide an environment as attractive thermodynamically as the micelles from which the lipids came. Polar moieties can reside in water-swollen hydrophilic domains, and their respective hydrocarbon moieties will reside in adjacent oleophilic ones. Once in this situation, a polar lipid will diffuse along an interface driven by a concentration gradient.

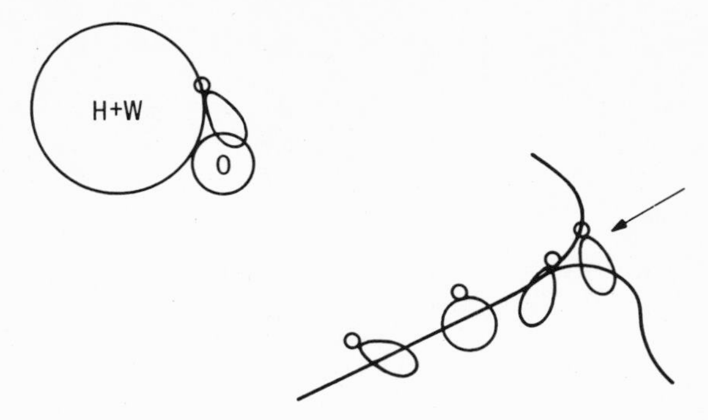

Fig. 12. Representation of the second and third steps in the process which occurs when a micellar lipid solution is contacted with a mixed-chain polymer. As in Figure 10, micelles that come in contact with oleophilic domains disrupt, and the lipids spread out on that surface. Lipid molecules are then able to migrate into the polymer by organizing themselves individually at the interfaces between adjoining oleophilic and hydrophilic domains.

This model has not yet been studied quantitatively, but it is easy to see that results something like those shown in Figures 1 and 2 might be expected. Providing the hydrophilic and oleophilic domains are generally nearly the same size, the number and length of continuous paths (made up of connecting interfaces; see Figure 8) starting at the surface will be highest when the volumetric fractions of the two are close to being equal. These paths would be the location of absorbed polar lipids. Diffusion along such paths of molecules with the size of bile lipids would be slow. Of the eleven conclusions stated above, number 9, enhancement of water absorption by the presence of lipids, is the only one which cannot be inferred from the model. The resistance of a sizeable fraction of absorbed lipids to leaching, number 11, is not thoroughly clear. However, some interpretation of this phenomenon may be obtained by considering the lipids with low polarity, such as cholesterol. Once in the polymer and away from the influence of micelles, there is little force to prevent them from diffusing away from the hydrophilic-oleophilic interfaces into the bulk of oleophilic domains. The trouble with this explanation is that the amount of cholesterol present in the biles studies was much too low to account for the apparently non-leachable one-third. There is, however, another possible source of low-polar lipid, lecithin. This phospholipid has both a positive and a negative ion. They can self neutralize in pairs and thus lose most of their polarity.

In summary, interesting results were obtained in the study of the interactions of micellar bile solutions with a new type of polymer, containing both hydrophilic and oleophilic chains. These results were interpreted in terms of a qualitative model. The model was based on a hypothesis which states that regional compatibility characteristics of both lipid and polymer are much more influential than averaged bulk characteristics in determining the nature of such interactions. Good correspondence was obtained between the model and the results from this limited investigation. Quantitative evaluation of the correlation must be left to a later time. Also left for later investigation is the question of how general the model is. The applicability of the model to other similar polymers and other micellar systems must be determined. To the extent that the model is general, it should be useful in the application of polymers to biomedical problems other than the one which stimulated the present work.

REFERENCES

1. H. E. Marsh, Jr., JPL Quarterly Tech. Rev., 1 (1), 49 (1971).
2. H. E. Marsh, Jr., and C. J. Wallace, JPL Quarterly Tech. Rev. 2 (4), 1 (1973).
3. D. J. Lyman, B. H. Loo, and R. W. Crawford, Biochemistry, 3 (7), 985 (1964).
4. D. J. Lyman, B. H. Loo, and W. M. Muir, Trans. Amer. Soc. Artif. Int. Organs, 11, 91 (1965).
5. D. J. Lyman, and B. H. Loo, J. Biomed. Mater. Res. 1, 17 (1967).
6. D. M. Small, M. C. Bourges, and D. G. Dervichian, Biochem. Biophys. Acta., 125, 563 (1966).
7. B. Vollmert, and H. Stutz, in Colloidal and Morphological Behavior of Block and Graft Copolymers (G. E. Molar, ed.), Plenum Press, New York-London, 1971, p. 209-221.

ANTICOAGULANT ACTIVITY OF SULFONATE POLYMERS AND COPOLYMERS

Harry P. Gregor

Department of Chemical Engineering and
Applied Chemistry
Columbia University, New York, New York 10027

INTRODUCTION

Heparin is a natural component of blood and one of the important compounds involved in natural processes of coagulation and anticoagulation. It is a polysaccharide having a molecular weight of about 12,000 and characterized by a high, negative fixed charge due to sulfonate and sulfate groups attached to the mixed copolymer of glucosamine and uronic acid. The chain contains also a high concentration of fixed carboxylic groups. An average of one sulfonate or sulfate group is found for each chain unit along with two carboxylic groups for each three chain units, based upon current evidence of the structure of this polymer. Being a highly charged polyacid, heparin is coagulated or precipitated by basic polyelectrolytes; under physiological conditions the polybasic polypeptide protamine is employed for this purpose. Heparin is the only biocolloid present in the human which does possess the high fixed sulfonate charge (1).

Heparin has been employed rather extensively for the purpose of imparting non-thrombogenicity to the surface of plastics by coulombic attachment to fixed basic groups, by direct chemical bonding to reactive groups on the surface of the plastic and by occlusion into a microporous plastic material from which it is diffused slowly. The literature on synthetic heparinoids is also extensive. A variety of sulfonate polymers ranging from poly(styrenesulfonic) acid to a number of sulfonated or sulfated natural products (including sulfated chitin) have been examined. Invariably, these materials have demonstrated limited effectiveness and often a high toxicity. Similarly, surfaces to which polystyrene sulfonic acid has been attached have not shown promise for non-thrombogenicity and there are reports of red blood cell damage by their use (2).

The sulfonate fixed charge is unique because this anion does not bind divalent cations (such as calcium and magnesium) as do carboxylic or phosphorous acid groups. For example, the common carboxylic polymer polyacrylic acid binds calcium strongly, forming a series of complexes with a calcium ion bound either to two adjacent carboxyl groups to form a chelate complex or singly to form an effective cross-link between colloids or polymers containing the carboxylic acid group. Polyacrylic acid is readily precipitated by the addition of small amounts of calcium, even at high concentrations of sodium chloride. On the other hand, a typical sulfonate polymer such as poly(styrenesulfonic acid) (PSSA) is not precipitated by calcium; indeed, even the barium salts of this polymer are soluble unless the polymer is of unusually high moleular weight, in which case a gel forms (3). It is for these and other reasons that it has been postulated that the effectiveness of heparin arises from its high negative fixed sulfonate charge and its ability to maintain this charge in the presence of calcium and magnesium.

Polymers prepared from the monomer vinylsulfonic acid are of particular interest; they are postulated to possess marked advantages over poly(styrenesulfonic) acid and to be more similar to heparin because they do not possess the aromatic ring to which the sulfonate is attached as in PSSA. There is a considerable body (4) of evidence in the ion-exchange polymer literature that the aromatic ring has a high degree of hydrophobic character even when sulfonated. For example, one finds the selective absorption of apolar cations even with fully sulfonated polystyrene. Poly(vinylsulfonic) acid has been reported to inhibit clotting and is used as an additive for bandages for that purpose (5).

This study reports on the screening of a number of sulfonate polymers and copolymers as synthetic heparinoids by standard tests for heparinoid activity.

EXPERIMENTAL

A group of homopolymers and copolymers of vinylsulfonic acid were prepared using conventional polymerization techniques. Their mole fraction of sulfonate groups present was determined by direct titration which is accurate to within 0.5 mole percent when carried out under conditions of high ionic strength. Polymers of polystyrene sulfonic acid of two molecular weights were also prepared as was the homopolymer of maleic anhydride, for purposes of comparison. Each of these polymers were characterized by as to their molecular weight by conventional light scattering techniques. The polymers were placed in the Na state at pH 7.4, dialyzed free of excess salt and freeze dried to obtain the dry powder. Table I summarizes the properties of these polymers. As used, the counter ion of the sulfonate groups was Na, the dicarboxylic itaconate was about half neutralized by Na as was the maleic acid, while acrylic and methacrylic acid groups were almost fully neutralized. The acrylamide groups were not measureably hydrolyzed by the procedures employed (6,7).

Two tests of the anticoagulant properties of these compounds were carried out. H. L. Nossel employed the following procedure: 10 mg of each sample was added to 5 ml of saline to form a solution which could be diluted further from 2 mg/ml. Then 0.1 ml of plasma (23 June 1970, M.D.) and 0.1 ml of the sample (or saline blank) were warmed to 37° for 30 sec and 0.1 ml thrombin solution (about 0.8 NIH units) was added and the clotting time measured.

The other procedure by O.D. Ratnoff employed one vol of a solution of the compound in barbital saline buffer to which was added 2 vol of normal platelet deficient plasma, the solution incubated at 37° for 1 minute, then 1 vol of bovine thrombin (Parke-Davis topical thrombin) at a concentration of 10 units/ml of barbital saline buffer was added and the clotting time was measured by tilting the tubes continuously. Table II lists clotting times (average of duplicate tests) for different concentrations of the compounds studied, all compared with heparin. The % heparin activity was calculated from a plot of clotting time vs. heparin concentration (taking 130 units as 1 mg heparin). A 2% heparin activity means that 2 mg of heparin gave the same clotting time as 100 mg of the compound, etc.

TABLE I

PROPERTIES OF POLYMERS

Cpd	Monomers	Sulfonate mole %	MW
A	VSA, Ita	42	50,000
B	VSA, Mal An	48	50,000
C	VSA, MA	11	100,000
D	VSA, AA	23	100,000
E	VSA, Ita	8	70,000
F	VSA, Am	25	400,000
G	VSA, Mal An	95	50,000
H	VSA	100	19,000
I	VSA	100	28,000
J	PSSA	100	525,000
K	PSSA	100	40,000
L	Mal An	0	10,000

VSA - vinylsulfonic acid; Ita-itaconic acid, Mal An - maleic anhydride, MA - methacrylic acid, AA - acrylic acid, Am - acrylamide.

DISCUSSION

Both procedures gave quite similar percentages of heparinoid activity on the part of the synthetic polymers, in terms of both relative and absolute values.

The VSA homopolymers (H,I) showed substantial anticlotting activity (14-16%). The PSSA polymers (J,K) were also reasonably active(5-12%), even though the molecular weight range here was considerable. The mixed sulfonate-carboxylate copolymers showed variable results, some of which were apparently inconsistent. For example, the 50% mole fraction VSA copolymers with the dicarboxylic itaconate and maleate moieties (A,B) were poor (0.01 - 0.04%) but the 8% sulfonate - itaconate polymer (E) was good (12,13%). Polymer G was 95% sulfonate so it should have been comparable to H and I, which it was considering the differences in MW. The methacrylic acid copolymer with VSA (C) was poorer (2,4%) than was the copolymer (D) with acrylic acid (8,12%) but the latter had twice the sulfonate content.

TABLE II

CLOTTING TIMES AND HEDARINOID ACTIVITY OF SYNTHETIC POLYMERS

	H.L.N. Procedure			O.D.R. Procedure		
Cpd	Conc mg/me	t sec	Heparin Activity %	conc mg/me	t sec	Heparin Activity %
A	2000	60.1	0.02	1000	37	0.04
B	2000	37.7	0.01	1000	28	0.02
C	20	74.0	2	10	37	4
D	2	32.4	12	2.5	27	8
E	2	34.5	13	2.5	31	12
F	20	25.7	1	250	40	0.2
G	2	29.2	12	10	57	6
H	2	45.4	16	2.5	35	14
I	2	42.0	15	2.5	40	16
J	2	32.5	12	2.5	28	9
K	2	24.9	10	10	49	5
L	2	27.1	11	40	59	2
heparin	0.2	23.5		0.08	22	
	0.3	44.5		0.15	25	
	.4	63.3		0.3	27	
	.5	>120		0.6	46	
Saline	.0	18.5		0.0	21	

The VSA copolymer with the only non-ionic comonomer acrylamide (F) was poor (0.2,1%) but it was also of very high MW. The only non-sulfonate copolymer (L) showed a rather different activity by the two procedures (2,11%). This material differs from other carboxylate polymers in that the first of the two groups is quite strong as an acid, while it trans-configuration makes it poorer at binding divalent metallic cations; chelate rings are not easily formed except at high pH levels of about 9-10.

The results reported herein must be regarded as quite preliminary. There is encouragement to be drawn from the high activity of the vinyl sulfonate polymers. Since these can be prepared readily and fractions of different molecular weights can be isolated easily by gel permeation chromatography, further studies are indicated.

The author acknowledges the support of the National Institute for Arthritis and Metabolic Diseases, Artificial Kidney Program. He wishes to thank H. L. Nossel, Department of Medicine, College of Physicians and Surgeons, Columbia University and O. D. Ratnoff, Department of Medicine, Lakeside Hospital, Case Western Reserve University for their generous and valued assistance.

REFERENCES

1. R. D. Rosenberg and P. S. Damus, J. Biol. Chem. 248, 6490 (1973).

2. R. I. Leininger in Biophysical Mechanisms in Vacular Homeostasis and Intravascular Thrombosis, P. N. Sawyer, Ed., Appleton-Centry-Crofts, N.Y., 1965.

3. M. H. Waxman, B. R. Sundheim and H. P. Gregor, J. Phys. Chem., 57, 969 (1953).

4. H. P. Gregor in Polyelectrolytes, E. Selegny, Ed., Reidel, Holland, 1974.

5. See egs. H. Tiedemann et al., Science, 164 (1969).

6. H. P. Gregor and F. C. Chlanda to be published.

7. An excellent general reference is S. D. Bruck, Blood Compatible Synthetic Polymers, Thomas, Springfield, Ill., 1974.

BOUND FRACTION MEASUREMENTS OF ADSORBED BLOOD PROTEINS

Bruce W. Morrissey and Robert R. Stromberg

Institute for Materials Research
National Bureau of Standards
Washington, D. C. 20234

An understanding of the conformation and conformational changes of proteins and enzymes concomitant upon adsorption is important in many areas of biochemistry. Clinical and commercial applications of matrix insolubilized enzymes (1), oxidative phosphorylation within mitochondria (2), the promotion of aggregation and adhesion of cells by glyco-proteins on the cell surface (3), and surface-induced blood coagulation (4-6) are examples of phenomena dependent on the conformation of proteins at interfaces. The possible effects of a given surface on a protein mixture would include, among others, permanent or reversible adsorption with or without denaturation or conformational changes, preferential adsorption of specific proteins, and changes in the microenvironment of enzymes.

A variety of techniques and materials have been utilized for the determination of the conformation of adsorbed proteins. Comparisons of the adsorbed molecular area and solution dimensions have been made by Bull (7) for egg albumin using the film balance, and by Brash and Lyman (8) for numerous serum proteins using internal reflection spectroscopy. Perturbations in the fluorescence spectrum resulting from adsorption were used by Katchalski (9) to demonstrate conformational changes of trypsin and chymotrypsin using a transmission technique, and by Loeb and Harrick (10) in conjunction with internal reflection spectroscopy to determine that serum albumin is apparently native while adsorbed. Kochwa and coworkers (11) applied potentiometric titrations to show that γG-globulin unfolds upon adsorption to a polystyrene latex under conditions of low surface coverage and acquires an antigenicity similar to heat denatured material. Ellipsometry has been utilized by Vroman (12) to qualitatively study sequential, competitive adsorption of blood proteins, while Smith et. al. (13) have employed the technique quantitatively to determine the adsorbed amounts and extensions of individual blood proteins. While the results obtained at the air-solution interface are consistent in generally showing that large changes in protein conformation may occur, there is no consensus regarding the conformation at the solid-solution interface.

In this paper* we report the results of the direct measurement of the number of attachments made by the carbonyl group of adsorbed blood proteins with a well characterized silica surface. These results are viewed as providing essential information on the conformation and changes in conformation of the adsorbing molecule. These studies have been carried out *in situ* on individual proteins as a function of the amount adsorbed, time of adsorption, pD, and ionic strength. Details of the interaction of blood proteins with surfaces are essential to both an understanding of surface-induced coagulation and the development of a rational approach to the selection of thromboresistant materials.

TECHNIQUES

The interaction of the chromophores of an adsorbed molecule with a surface frequently results in a shift of their characteristic spectral absorption bands. Typically, for the protein studies presented here, a shift of -20 cm^{-1} of the amide I band for free and bound carbonyl groups was observed. If these bands were well separated, one could immediately use the optical density of the shifted band to determine the bound fraction p, the fraction of the carbonyl chromophores of an adsorbed protein molecule directly in contact with the surface. Knowledge of the amino acid composition of the protein enables a calculation of the actual number of carbonyl attachments per molecule.

---

*A more complete description of the work presented here has been published, Colloid Interface Sci., 46, 152(1974).

Usually, however, there is a significant overlap of the two bands and difference techniques must be used. Fontana and Thomas (14) originally described a method whereby a polymer is adsorbed from solution on a high surface-area powder, the suspension centrifuged, and an infrared difference spectrum recorded for the resulting gel. This technique was modified (15) to permit a direct analysis of the suspension, thereby removing uncertainties in polymer concentration and has been used by a number of investigators (16-18) to study the conformation of synthetic polymers adsorbed from organic solvents.

For the determination of the bound fraction for an adsorbed protein, it is necessary to use D_2O solutions of deuterated proteins to obtain a window in the 1650 cm^{-1} region. An analysis, based on Beer's law and utilizing the optical density of the unshifted peak, was used to resolve the difference spectrum arising from a protein solution-silica suspension in the sample beam versus the same protein solution in the reference beam.

EXPERIMENTAL

Bovine serum albumin (4x crystallized), bovine fibrinogen (Cohn fraction I, 60% clottable), and bovine prothrombin (fraction III-2) were obtained from Nutritional Biochemicals.** The serum albumin was deuterated (19), lyophilized, and stored in vacuo at 4°C. The prothrombin was dissolved in 0.1M D_2O phosphate buffer pD 7.4, dialyzed overnight against buffer, and filtered through a well washed 0.8 μm pore size filter just prior to use. Fibrinogen was purified by the method of Laki(20). All protein solutions were prepared using 0.1M D_2O phosphate buffer. Adjustments of pD were made with DCl or NaOD while ionic strength adjustments were carried out with KCl.

**Certain commercial equipment and instruments are identified here and elsewhere in this publication in order to specify adequately the experimental procedure. In no case does such identification imply recommendation or endorsement by the National Bureau of Standards, nor does it imply that the equipment or instruments identified are necessarily the best available for the purpose.

The adsorbent used for all experiments was a fumed, non-porous silica with a nominal particle size of 0.012 μm (Cab-O-Sil M-5, Cabot Corp.). The adsorbent was heated in vacuo at 110°C overnight and stored over silica gel just prior to use. The surface area, as determined by BET N_2 analysis, was found to be 204 ± 20 m^2/gm.

The adsorption isotherms were constructed at 23.5 ± 0.5°C by shaking known amounts of silica and protein solution for three hours, at which time the silica was removed by filtration through a 0.8 μm filter. The amount of adsorbed protein was determined from the protein concentration changes measured spectrophotometrically at 280 nm. In a number of cases, the ratio of silica to solution volume was varied with no effect on the protein adsorption isotherms. For studies of adsorbance as a function of time, the quantities of silica and protein solution were suitably scaled up and aliquots withdrawn and filtered with time.

For bound fraction studies, 0.029gm silica per ml protein solution was found generally to give stable suspensions suitable for infrared analysis. The difference spectrum of the protein-silica suspension versus the same protein solution was obtained for the region 1750-1550 cm^{-1} using matched 0.1 mm CaF_2 cells at 10x ordinate expansion. The extinction coefficient of the adsorbed carbonyl group was obtained by scanning successive dilutions of the reference solution against the protein-silica suspension and noting the dilution at which the peak due to the unbound chromophores disappeared. Each reported bound fraction value represents the average of from two to four infrared scans on the same protein suspension.

RESULTS AND DISCUSSION

The isotherms of serum albumin and prothrombin are given in Figure 1 along with values of the bound fraction. The bound fraction of fibrinogen was found in general to depend on the equilibrium concentration (and hence the adsorbance). In Figure 2 are shown the results of a set of four experiments on fibrinogen, each of which utilized purifed protein to prepare solutions with a range of concentrations. Portions of the fibrinogen isotherm were inaccessible for bound fraction measurements using the technique described

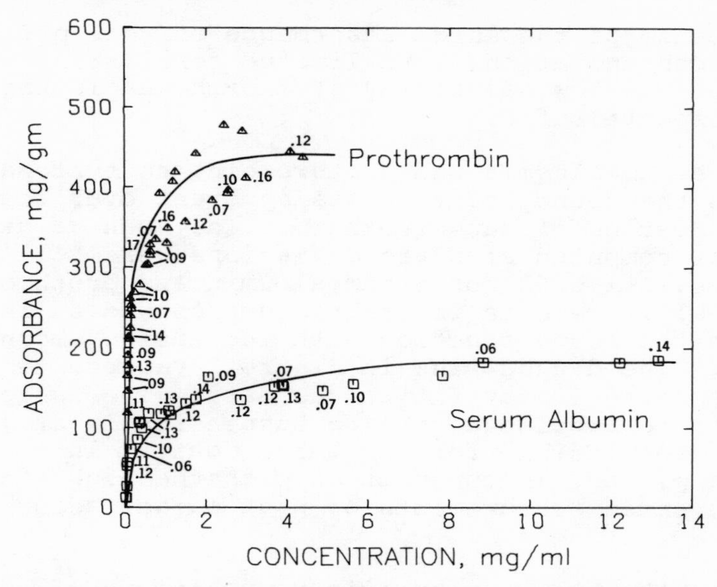

Figure 1: Adsorption isotherm for bovine serum albumin and prothrombin at pD 7.4 on silica. Values of bound fraction given for isotherm points.

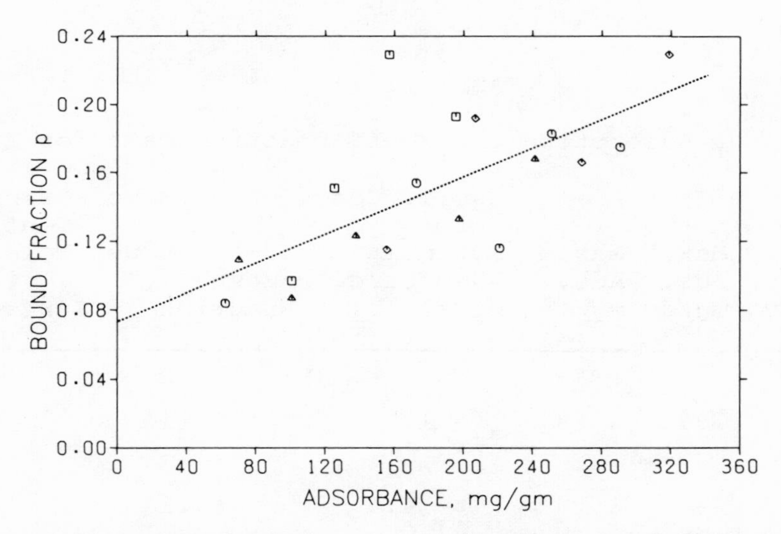

Figure 2: Bound fraction vs. adsorbance of bovine fibrinogen for four different experiments. Each symbol represents a separate run. Adsorption at pD 7.4 on silica.

above because of the large adsorbance of this protein (initial concentrations were limited for fibrinogen because of the low solubility of fibrinogen in the D_2O - phosphate buffer).

For serum albumin and prothrombin, no systematic change in the bound fraction was apparent over the measured portion of each isotherm. The mean values with their computed standard deviations were 0.11 ± 0.02 and 0.11 ± 0.03 for serum albumin and prothrombin, respectively. Results for fibrinogen showed an increase in the bound fraction with increasing amount adsorbed. The dashed line in Figure 2 represents a least-squares linear fit to the data. The unstable nature of the fibrinogen-silica suspension is at least partially responsible for the large scatter in the values of p, and, in conjunction with the high adsorbance, prevented measurements of p at larger adsorbance values.

Using the known amino acid compositions of serum albumin, prothrombin, and fibrinogen, the bound fractions reported translate into 77 carbonyl contacts for serum albumin, 80 for prothrombin, and 176-703 for fibrinogen with the silica surface as shown in Table I.

TABLE I

Summary of Adsorption and Bound Fraction Data for Proteins

Protein	Max. Ads. mg/gm	Max. Ads. mg/m^2	Equil. Conc. for Maximum Adsorbance mg/ml	Bound Fraction	No. Carbonyl Contacts per Molecule with Surface
Serum Albumin	180	0.88	11	0.11	77
Prothrombin	440	2.16	4.0	0.11	80
Fibrinogen	890	4.36	1.0	0.05-0.20	176-703

Despite this extensive interaction with the surface, the results imply that the adsorbed protein molecules are still relatively unchanged by the surface, compared to synthetic random coil polymers. Adsorption of poly(ethylene o-phthalate) (17), poly(methyl methacrylate) (16), polystyrene (18), and poly(4-vinyl pyridine) (15) on Cab-O-Sil gave bound fractions of 0.37, 0.35, 0.24, and 0.35, respectively. In each of these cases the synthetic molecules appear to lie close to the surface in a presumably spread condition.

The bound fraction data can yield information on changes in the conformation of the adsorbed protein molecule as a result of increasing competition for surface sites. A decrease in the bound fraction values with increasing surface population would be expected if changes in conformation were to occur. Such changes in conformation of synthetic random coil polymers have been observed (15, 16, 21). It has been argued (22) that an adsorbed serum albumin molecule should initially unfold, thereby increasing the value of p, which would presumably then decrease with increasing surface population. Our results for serum albumin and prothrombin, Figure 1, show that the bound fraction remains constant, indicating that the conformation does not change as the surface population increases. The relatively low value of p for these proteins as compared to the synthetic polymers, and the constant value of p over the isotherms, strongly suggests that the internal bonding and disulfide crosslinks of these globular proteins are sufficient to preserve the basic tertiary structure of the molecules.

The adsorption behavior of fibrinogen (Fig. 2), which shows a direct dependence of bound fraction on adsorbance, does not necessarily represent a conformational change. As noted above, previous experience with synthetic polymers has shown that conformation changes resulting from increases in surface population are accompanied by decreases in the bound fraction. The bovine fibrinogen molecule is rodlike with a length of between 400-600 AU (23) and can easily interact with more than one silica particle. It has been postulated (24) that the solubility of fibrinogen is due to electrostatic repulsion between regions of high negative charge density localized in fibrinopeptides A and B. If adsorption and interaction with the hydrophilic surface were to effectively neutralize this repulsion, fibrinogen molecules adsorbed to different silica particles could interact like fibrin, thereby increasing the surface attachments of each fibrinogen molecule. Such an interfacial aggregation (22) could result in an increase in bound fraction values with increasing adsorbance, without significantly hindering further adsorption to the remaining available surface.

Our data for maximum adsorbance can be converted to amounts/unit area using the measured surface area. These results are presented in Table I and agree quite favorably with those found for adsorption of human blood proteins on quartz using ellipsometric techniques (25). Results obtained by MacRitchie (22) for adsorption of bovine serum albumin on a silica surface (also 200 m^2/gm) at pH 7.5 were only one third as great as that found in the present study, while adsorption on flat polymer surfaces determined using internal reflection techniques (8) gave generally higher amounts. The reasons for these differences are not obvious.

The possibility that the conformation of the adsorbed protein might vary as a function of the time of adsorption was tested by determining the bound fraction for each protein periodically following the formation of a suspension. Figure 3 illustrates some typical results obtained for serum albumin, prothrombin, and fibrinogen where the bound fraction values are listed over each curve at the time of measurement. Although the rates of adsorption cannot be compared since the initial conditions are different for each experiment, it is clear that there is no variation in the bound fraction for any of the proteins over the time period investigated at these conditions. This indicates that molecular rearrangements do not occur following adsorption at the concentrations studied. Although the adsorption rates for prothrombin and fibrinogen were extremely high as a result of the conditions required for bound fraction measurements, the relatively slow adsorption of serum albumin allows measurements of p as a function of time which shows that the conformation is not dependent upon the surface population, as was found in the equilibrium adsorption studies (Fig. 1). The slight decrease in adsorbance of prothrombin with time shown in Fig. 3 is possibly caused by a dissociation at the low final solution concentration. Prothrombin has been reported (26) to associate reversibly at low ionic strengths.

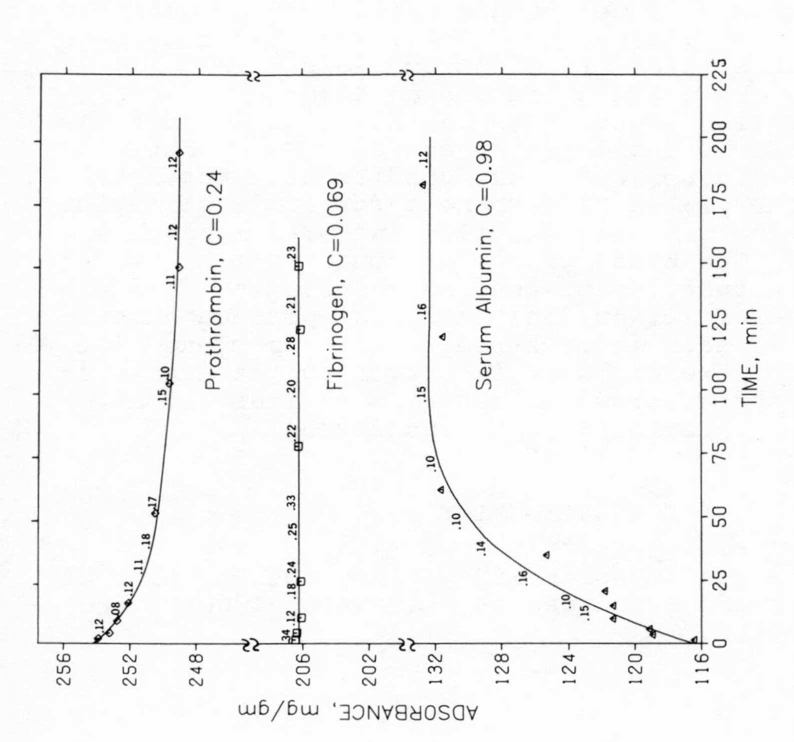

Figure 3: Adsorbance vs. time for bovine serum albumin, prothrombin, and fibrinogen. Values of bound fraction are given over each curve at the time measured. Adsorption on silica at pD 7.4

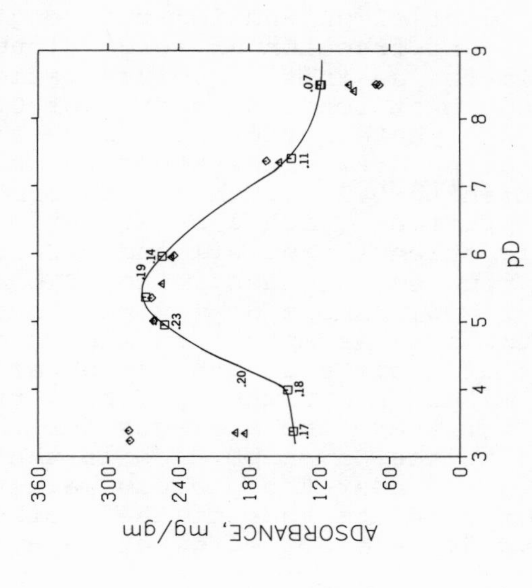

Figure 4: Adsorbance vs. pD for bovine serum albumin at 0.1 (□), 0.2 (△), and 0.5 (◇) ionic strength on silica (0.014 gm/ml solution). Values of bound fraction given over curve for 0.1 I at the pD measured.

Experiments have also been carried out to ascertain the effect of pD and ionic strength on the adsorbance and bound fraction of serum albumin. The results obtained for an initial concentration of 4.3 mg/ml serum albumin at ionic strengths of 0.1, 0.2, and 0.5 in D_2O-phosphate buffer are presented in Figure 4. The curve drawn represents the effect of pD for ionic strength 0.1. Both the adsorbance and bound fraction profiles reach a broad maximum near pD 5.2 for ionic strengths of 0.1 and 0.2 reminiscent of Bull's (27) studies on glass. The pD profile of adsorbance for I = 0.5 shows no maximum, although there appears to be an asymptote in the acid region. Since no effect of ionic strength was observed on the bound fraction within experimental error, the values of p given in Fig. 4 are the averages for the three ionic strengths at the given pD. While the adsorbance values at pD 7.4 are near the plateau values, those at pD 5.0 are only on the knee of the isotherm (plateau adsorbance at pD 5: A = 308 mg/gm at C_{equil} = 3.0 mg/ml).

Relating the plateau values of protein isotherms for a solid-liquid system with a close-packed monolayer can be very misleading. Using accepted figures for the molecular geometry of serum albumin in solution (8), one would calculate an adsorbance for our system of 520 - 1460 mg/gm depending on whether the molecule is adsorbed side-on or end-on. We do not approach these values under any conditions. Further, at pD 3.35, the plateau adsorbances for ionic strengths of 0.1, 0.2, and 0.5 are 142, 193, and $\sim$365 mg/gm, respectively. The bound fraction (conformation) is, however, not a function of the ionic strength (Fig. 4). One would, therefore, predict identical plateau adsorbance values for each ionic strength. It is clear that caution must be exercised in interpreting the available surface area per adsorbed molecule as an indication of changes in conformation upon adsorption.

ACKNOWLEDGMENT

This work was supported in part by the Division of Blood Diseases and Resources of the National Heart and Lung Institute.

REFERENCES

1. Mosbach, K., Sci. Amer., 224 (3), 26 (1971).
2. Green, D. E., and Young, J. H., Amer. Sci., 59, 92 (1971).
3. Pessac, B., and Defendi, V., Science, 175, 898 (1972).
4. Hulbert, S. F., King, F. M., and Klawitter, J. J., J. Biomed. Mater. Res. Symp., No. 2 (Part 1), 69 (1971).
5. Bruck, S. D., Biomaterials, Med. Devices and Artificial Organs 1, 79 (1973).
6. Brash, J. L., and Lyman, D. J. in "The Chemistry of Biosurfaces" (M. L. Hair, Ed.) 1, p. 177. Marcel Dekker, Inc., New York, 1971.
7. Bull, H. B., J. Colloid Interface Sci., 41, 305 (1972).
8. Brash, J. L. and Lyman, D. J., J. Biomed. Mater. Res., 3, 175 (1969).
9. Gabel, D., Steinberg, I. Z., and Katchalski, E., Biochem. 10, 4661 (1971).
10. Loeb, G. I. and Harrick, N. J., Anal. Chem., 45, 687 (1973).
11. Kochwa, S., Brownell, M., Rosenfield, R. E., Wasserman, L. R., J. Immun., 99, 981 (1967).
12. Vroman, L. and Adams, A. L., Thrombos. Diathes. Haemorrh., 18, 510 (1967).
13. Smith, L. E., Fenstermaker, C. A., and Stromberg, R. R., ACS Polymer Preprints, 11, 1376 (1970).
14. Fontana, B. J. and Thomas, J. R., J. Phys. Chem., 65, 480 (1961).
15. Thies, C., Peyser., and Ullman, R., Proceedings of the 4th International Congress on Surface Activity, Brussels, 1964, Vol. 2, Gordon and Breach, N.Y. 1967, p. 1041.
16. Thies, C., J. Phys. Chem., 70, 3783 (1966).
17. Peyser, P., Tutas, D. J., and Stromberg, R. R., J. Poly. Sci. A-1, 5, 651 (1967).
18. Herd, J. M., Hopkins, A. J., and Howard, G. J., J. Poly. Sci. Part C, 34, 211 (1971).
19. Susi, H., Timasheff, S. N., and Stevens, L., J. Biol. Chem., 242, 5460 (1967).
20. Laki, K., Arch. Biochem. Biophys. 32, 317 (1951).
21. Stromberg, R. R., Tutas, D. J., and Passaglia, E., J. Phys. Chem., 69, 3955 (1965).
22. MacRitchie, F., J. Colloid Interface Sci., 38, 484 (1972).
23. Laki, K., "Fibrinogen," p. 4. Marcel Dekker, Inc., New York, 1968.

24. Laki, K., Op. Cit. p. 6.
25. Smith. L. E., Fenstermaker, C. A., and Stromberg, R. R., to be published.
26. Tishkoff, G. H., Williams, L. C., and Brown, D. M., J. Biol. Chem., 243, 451 (1968).
27. Bull, H. B., Arch. Biochem. Biophys., 68, 102 (1957).

COMPETITIVE ADSORPTION OF PLASMA PROTEINS ONTO POLYMER SURFACES

S.W. Kim, R.G. Lee, C. Adamson & D.J. Lyman

Division of Materials Science & Engineering
College of Engineering
University of Utah
Salt Lake City, Utah 84112

INTRODUCTION

Several investigators have found that foreign surfaces, when exposed to blood, adsorb plasma proteins. (1,2,3) Platelet adhesion to this proteinated surface is the first observable event occurring in clotting on foreign surfaces. It has been demonstrated that hydrophobic polymers that had been pre-coated with albumin showed essentially no platelet adhesion when tested in an ex vivo cell (4); on the contrary, fibrinogen or γ-globulin coated surfaces show more platelet adhesion and release of constituents. (5) Our previous adsorption study showed that a relatively non-thrombogenic surface has a faster rate of albumin adsorption with greater concentration as compared with other proteins than does a thrombogenic material. (6,7) If we consider the proteinated surfaces as an acceptor in a platelet adhesion mechanism, (8) the significance of the nature of adsorbed protein is in terms of its role in platelet adhesion. A possible mechanism of adhesion could involve enzyme-substrate complex formation between platelet glycoprotein glycosyl transferases and surface adsorbed glycoproteins such as γ-globulin, fibrinogen and prothrombin. (8) However, to interrelate these concepts, one must show that adsorption characteristics shown by individual proteins also hold during adsorption of mixtures. To do this, mixtures of albumin, γ-globulin, fibrinogen and prothrombin at their physiological solution concentration ratio (with one of the proteins labelled with I^{125} were studied to determine the surface composition of each adsorbed protein and their respective equilibrium time selected polymer surfaces.

EXPERIMENT

Proteins used in this study are albumin (bovine crystalline); fibrinogen (Cohn Fraction I); γ-globulin (bovine Fraction II); Albumin I^{125} tagged (Squibb, activity 10μc/mg); γ-globulin I^{125} and Fibrinogen I^{125} (New England Nuclear Company). The protein solutions were 50 mg albumin, 30 mg γ-globulin, and 15 mg fibrinogen in 200 ml buffered saline. The polymer films used in this study are polydimethyl siloxane (Silastic Rubber, medical grade non-reinforced, Dow Corning); fluorinated ethylene/propylene copolymer (Teflon, FEP, Type A, DuPont); and a segmented copolyether-urethane-urea,(9) based on polypropylene glycol mol. wt. 1025, methylene bis (4-phenylisocyanate) and 1,2 - ethanediamine. An adsorption cell similar to that designed by Lee and Andrade (10) is being used in this experiment (Fig. 1). This design avoids air water interface, controls the washing time to remove excess clinging solution, is convenient in manipulation and allows us to obtain kinetic adsorption data. A Nuclear Chicago 2 Channel liquid scintillation spectrometer with 50 sample capacity is used for determination of radioactivity on polymer surfaces.

RESULTS AND DISCUSSION

The adsorption kinetics of albumin from its aqueous mixture solution (pH = 7.4 buffer) onto surfaces of Silastic Rubber (SR), fluorinated ethylene propylene copolymer (FEP) and the block copolyether-urethane-urea (PEUU) were studied at 37°C. The kinetic data for the single protein adsorption were also obtained for these three surfaces. The rate of adsorption and the adsorbed amount of protein was found to be dependent on the chemical nature of the polymer and on the composition of the protein mixture (Fig. 2-4). The plateau time of albumin adsorption onto FEP is about

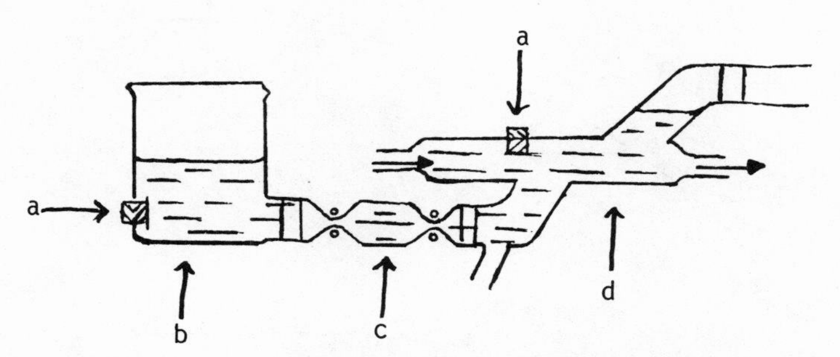

Figure 1. Schematic of protein adsorption cell system: (a) are magnets to move sample; (b) is adsorption cell; (c) is transfer lock; (d) is washing cell.

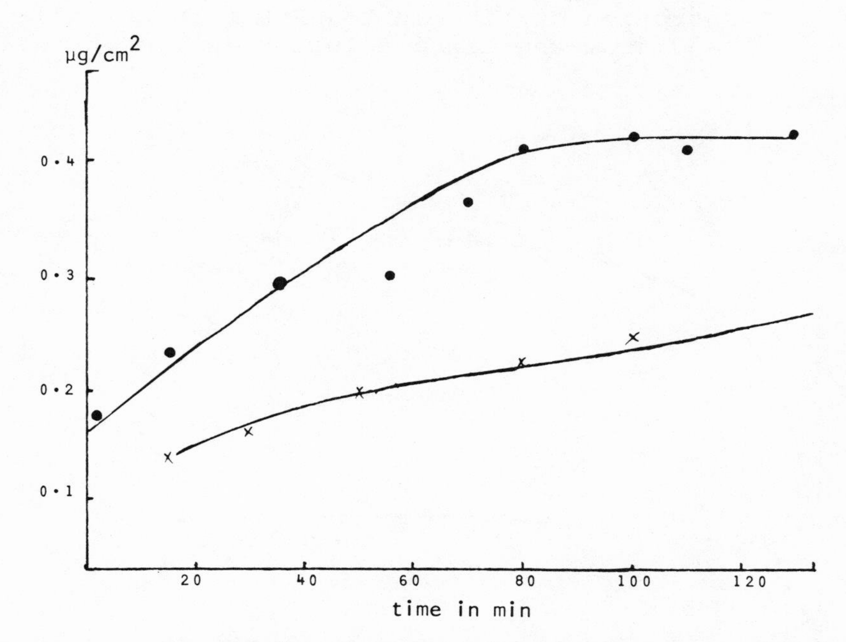

Figure 2. Adsorption of albumin to FEP from a single solution (•) and from the mixed solution (x).

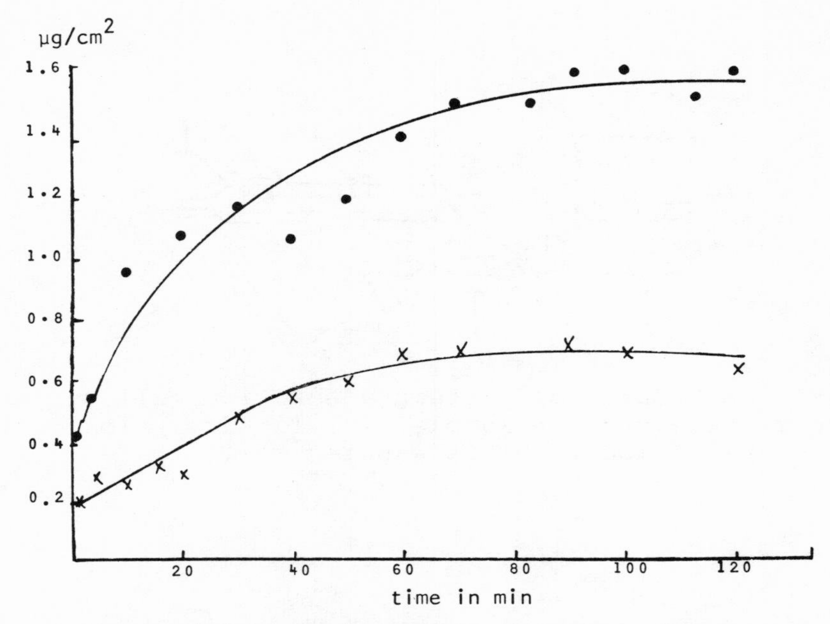

Figure 3. Adsorption of albumin to PEUU from a single solution (●) and from the mixed solution (x).

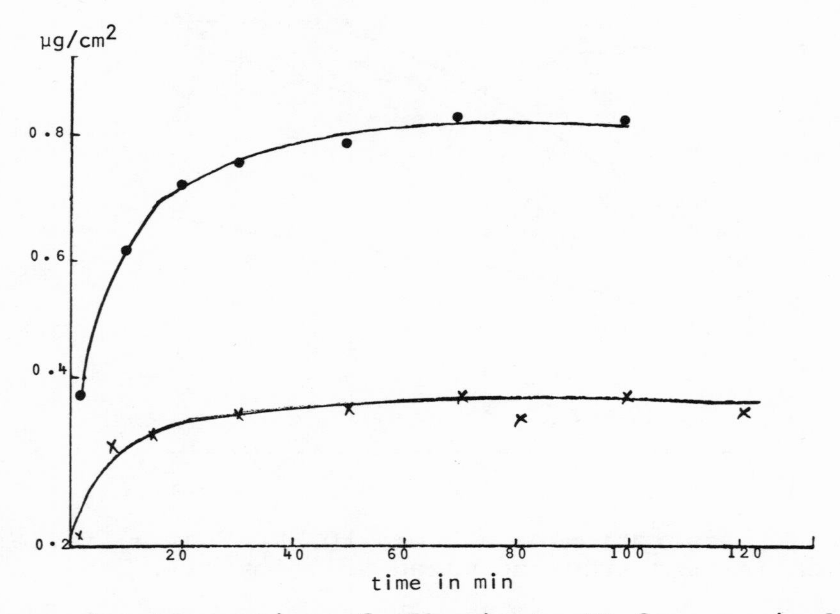

Figure 4. Adsorption of albumin to SR from a single solution (●) and from the mixed mixed solution (x).

90 minutes from the single protein solution and 150 minutes from the mixed protein solution (Fig. 2). The 60 minute difference is due to the competitive adsorption of γ-globulin and fibrinogen. In contrast, the non-thrombogenic PEUU shows similar albumin adsorption times for both single protein and mixed protein solutions (Fig. 3), indicating that the surface may have a larger amount of albumin on its surface. The adsorption pattern for SR is similar to PEUU although the total concentration of albumin on the surface is much less than the PEUU (Fig. 4). These experimental results (Table I) are very significant if we consider the platelet/acceptor complex model as discussed earlier. An albuminated surface cannot be an acceptor for platelets in the proposed model since albumin does not include any glycoprotein as an acceptor although fibrinogen and γ-globulin do. Therefore, it would appear that the chemical nature of the polymer surface determines the final composition of the adsorbed protein monolayer; and if the layer is primarily albumin, the polymer will be non-thrombogenic. If the surface has a large concentration of proteins such as γ-globulin and fibrinogen, the polymer will be thrombogenic.

TABLE I

PLATEAU TIME AND ADSORBED CONCENTRATIONS OF ALBUMIN ON POLYMER SURFACES

Surface*	Plateau Time, Min.		Surface Conc. μg/cm^2	
	Alb.	Alb. Mix.	Alb.	Alb. Mix.
FEP	90	150	0.42	0.27
SR	75	80	0.81	0.38
PEUU	70	70	1.6	0.7

*Polymer codes are given in text

ACKNOWLEDGMENT

This work was supported by a National Science Foundation Grant No. GK 29382. The authors wish to thank Dr. J. D. Andrade and Mr. H. B. Lee for their helpful suggestions.

REFERENCES

1. Brash, J. L. and Lyman, D. J., J. Biomed. Mat. Res., 3,175 (1969).
2. Baier, R. E. and Dutton, R. C., J. Biomed. Mat. Res., 3,191 (1969).
3. Scarborough, D.E., Mason, R.G., Dalldorf, F.C.T., Brinkhaus, K.M., Lab. Invest., 20,120,(1970).
4. Lyman, D. J., Klein, K. G., Brash, J. J. Fritzinger, B. K., Andrade, J. D., Bonomo, F.S., Thromb. Diath. Haem. (Suppl.), 42,109 (1970).
5. Lyman, D. J. and Metcalf, L. C., manuscript in preparation.
6. Kim, S.W. and Lee, R.G., J. Biomed. Mat. Res., (1974) in press.
7. Lyman, D.J. and Kim, S.W., Actualities Nephrologiques de L'Hopital Necker (1972) p. 97.
8. Lee, R.G., and Kim, S.W., J. Biomed. Mat. Res. (1974) in press.
9. Lyman, D.J., Kwan-Gett, C., Zwart, H. H. J., Bland, A., Eastwood, N., Kawai, J., Kolff, W. J., Trans. Amer. Soc. Art. Int. Organs, 17,456 (1971)
10. Andrade, J. D., J. Assoc. Advancement Med. Invest., 7,110 (1973) in press.

MECHANICALLY STABLE NON-THROMBOGENIC HYDROGELS

J. J. Kearney and I. Amara
Research Triangle Institute
Research Triangle Park, N.C.

N. B. McDevitt
University of North Carolina
Chapel Hill, Chapel Hill, N.C.

Hydrogels, especially those of polyacrylamide have been reported as showing good non-thrombogenicity by several investigators (1,2). These are highly hydrated materials, on the order of 50 to 200% hydration and, therefore, possess very little mechanical strength. Because of this, it has not been possible to utilize the excellent non-thrombogenic character of these surfaces to advantage.

The objective of our work was to synthesize a surface similar to that of a hydrogel on the surface of an inert substrate material which possesses the mechanical properties suitable for the projected end use of the material.

MATERIALS

The substrate material to which the surface grafts were made was Silicone Rubber. This was a product of Dow-Corning designated Silastic 601-445 (medical grade). This was a copolymer of dimethyl and methyl vinyl siloxane cross-linked through the vinyl groups. It was either used as received or rinsed with isotonic saline solution before use.

Acrylamide (Eastman 5521) was recrystallized three times from acetone before use.

N,N[1] -methylene-bisacrylamide (EASTMAN 8383) was used without further purification.

Cupric Chloride (Fisher C-456) and Ferrous Ammonium Sulfate (J. T. Baker 2054) were used as received.

BLOOD TESTING

In Vitro. All grafted materials were tested by exposing to whole human blood in the Mason-Ikenberry modification of the Lyman Cell or in the case of grafted tubes the inner surface of the tube is filled with whole human blood for thirty minutes.

The Mason-Ikenberry cell has been described elsewhere and this will not be repeated here. The grafted tube was used as a test vessel by clamping a Siliconized needle into each end of the tube. To one of these needles is attached a three-way polypropylene stopcock which is attached to a venipuncture needle. The tube which has been previously filled with saline is opened to the venous puncture needle and the saline displaced by blood using venous pressure as the driving force. The surfaces of interest are exposed to blood for five minutes in the M-I cell and for thirty minutes in the tubes. The blood is emptied into a centrifuge tube containing sodium citrate and spun to separate out the red cells. Partial Thromboplastin Times (PTT) are run on the plasma. The results are compared to PTT run on blood exposed to Siliconized Glass and ungrafted Silastic surfaces.

In Vivo. The grafted surfaces were evaluated in vivo by implanting a grafted tube 3 mm I.D. x 35 mm in length or 6 mm I.D. x 35 mm in length in the inferior vena cava of dogs of 20-25 kgs. These dogs were free of filarial worms with a hematocrit of at least 35. The flow was monitored continuously by either an electromagnetic flow probe placed around the vena cava at the downstream end of the tube or with a Doppler (ultra sonic) flow probe placed around the implanted tube. Mechanical Properties: Tensile strength, Modulus and % elongation were measured with an Instron Table Model TM-M.

GRAFTING

Silastic 601-445 (Dow-Corning) was immersed in a 10% solution of acrylamide in water and varying amounts of either ferrous ammonium sulfate or cupric chloride was added. The tube and contents was degassed on a high vacuum rack by the freeze thaw method to a pressure of 10^{-5}mm and sealed off. The end of the degassed tube was inserted in a rubber coupling attached to a Heller stirring motor and rotated about its long axis at about 60 RPM. The tube was positioned at a distance from a Cobalt source of 1500 Curies so as to absorb at the rate .001 MR/hr. It was possible at this distance to position the tube vertically so that the dose was fairly uniform from end to end and the resulting graft was uniformally distributed along the tube.

The concentration of transition metal salt that gave optimum graft with a minimum of homopolymerization was 10^{-2}M/L of cupric chloride.

N,N^{1} -methylene-bisacrylamide was used as the cross-linking agent in 10% of the monomer weight.

RESULTS

Acrylamide was grafted onto Silastic primarily on the surface of the substrate. The presence of the acrylamide on the surface was demonstrated by ATIR spectra of the surface which shows an acrylamide spectra overlaying a background of Silicone Rubber. The surface location of the graft is also demonstrated by immersing a solution of a grafted tube in a solution of Amido Black which dyes polyacrylamide blue. The sectioned tube appears under microscopic examination to consist of three concentric rings, the inner and outer rings 0.1 mm in thickness and blue in color and the center ring much thicker and clear.

The prevention of homopolymerization is necessary since gellation takes place when homopolymer is formed in solution and the resulting decrease in diffusion rate of the monomer to the substrate surface severely limits the amount of surface graft that can be achieved.

It has been noted (3) that transition metal ions will interact with radicals in solution and on the substrate surface at different rates so that by adjusting the ion concentrations, it is possible to inhibit one mode of polymerization more than the other. Fortunately, the homopolymerization is affected to a greater extent than the grafting.

Ferrous ion was found to inhibit homopolymerization more effectively than Cupric ion, however, the Ferrous ion is reduced to an iron compound which has a detrimental effect on PTT. Table 1 shows a comparison of PTT of acrylamide grafts made in solutions containing 3.3 x 10^{-1}M/L Fe^{++} ions. These tubes have yellow color that could not be washed out. It appears that the PTT decreases with increasing graft; this is misleading since the more highly grafted tubes seem to contain more absorbed iron compound than the less highly grafted tubes.

A comparison was made of Cu^{++} ion and Fe^{++} and this is summarized in Table II. The acrylamide grafts made in solutions containing cupric ion show better PTT values and have no color. The lower concentration of Cu^{++} ion is still sufficient to prevent homopolymerization and these grafts show a higher PTT. These values, while improved, did not come up to those obtained by grafting acrylamide to polyethylene film.

We suspected an incomplete removal of acrylamide monomer might influence the PTT. Tubes grafted from an acrylamide solution were washed for 48 hours, filled with isotonic saline solution for thirty minutes and the saline drained into a quartz cuvette. The solution was measured against saline solution in a ultra violet spectrophotometer. Adsorption at 210 mm indicated the presence of acrylamide in the wash water. The washing was continued for fourteen days measuring the adsorption of saline each day. At the end of this time, the concentration in the wash was down to 10^{-7}g/100 ml. A group of grafted tubes washed for ten days before testing gave the improved PTT values recorded in Table III.

We had no method of determining whether these surfaces were cross-linked although we would expect them to be because of the system of initiator being used. In order to gain some control over the degree of cross-linking, N,N^{1}-methylene-bisacrylamide was

TABLE I

% Graft	PTT sec	PTT_1*	PTT_2 ∓
2.5	257	82	130
3.4	201	71	112
4.8	clotted	-	-
7.3	162	52	82
7.3	171	46	73
7.6	154	42	66
9.3	144	46	73
9.5	147	40	64

* PTT relative to Silastic
∓ PTT relative to Siliconized Glass. These designations are the same for all the tables.

TABLE II

% Graft	FAS * M/L	$CuCl_2$ M/L	PTT sec	PTT_1	PTT_2
8.6	10^{-1}	-	clotted	-	-
9.3	10^{-1}	-	144	46	73
9.4	10^{-1}	-	69	21	
9.5	10^{-1}	-	147	40	64
0.5	-	10^{-3}	278	83	131
1.3	-	10^{-3}	277	82	130
0.6	-	10^{-2}	258	55	87
1.3	-	10^{-2}	253	54	85

* FAS - Ferrous Ammonium Sulfate

TABLE III

% Graft	PTT sec	PTT_1	PTT_2
2.0	370	121	191
5.0	361	118	186
8.1	360	118	186
10.5	339	115	180

TABLE IV

% X	% Graft	PTT sec	PTT_1	PTT_2
7	3.3	349	98	155
7	4.1	517	123	194
7	9.5	483	136	215
3	3.7	363	86	136
3	10.3	601	143	226

TABLE V

% Graft	Load to Break (Kg)	% Extension	Modulus (dynes/cm^2)
0	4.25	1720	3.13×10^7
0	4.2*	1748	3.78×10^7
0.32	4.5	1634	4.03×10^7
0.45	4.1*	1720	3.5×10^7
6.9	1.6	368	9.2×10^7
7.2	3.4*	1456	4.1×10^7
7.5ǂ	1.5	560	4.5×10^7
8.5ǂ	2.9*	1306	3.7×10^7

* these specimens tested under water

ǂ contains 5% cross-linker (MBA)

added to the solution. Table IV shows the improvement in PTT which resulted from the extended washing and addition of cross-linking agent. While there is a tendency for the PTT to improve with total graft, there is no analogous improvement in PTT with increased cross-linking agent.

Cross-linked acrylamide surface grafts were made to Silastic tubing in the form of vena cava tubes. These were 3 mm I.D. and were evaluated by implanting in the vena cava (6 mm I.D.) and in the Iliac branch of the Femoral vein (3 mm I.D.). In the latter site, a Silastic tube of similar dimensions was implanted simultaneously in the other Iliac branch for direct comparison. Both acrylamide grafted tubes remained open during the two hour test period, although the flow diminished in the vena cava tube. The tube in the Iliac vein maintained 100% flow throughout the test while the companion Silastic tube occluded after 43 minutes. Fig. 1 shows the relative flow vs. the time of flow and this is obtained by plotting the flow of time "t" divided by the flow at time "o" against the time.

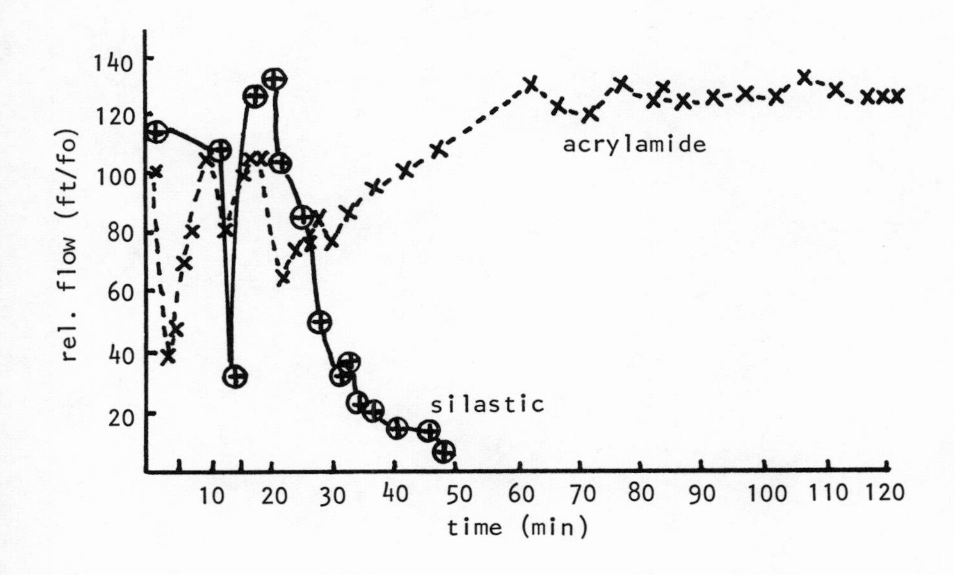

Figure 1

In order for this technique of producing non-thrombogenic surfaces to be useful, it must not result in any significant change in the mechanical properties of the substrate material. Table V is a summary of the properties of Silastic surface grafted with acrylamide. These specimens were tested in air and also in water. The latter testing was undertaken because the anticipated use of these materials is in an aqueous environment. When tested in the environment of anticipated use, the surface grafting has not altered the mechanically properties beyond the range of useability.

REFERENCES

1. B. D. Halpern, H. Cheng, S. Kuo, and H. Greenberg, Hydrogels and Non-Thrombogenic Surfaces, Proceedings, Artificial Heart Program, 1969.
2. F. Merril, Report, Fifth Annual Contractors Meeting, Artificial Kidney-Chronic Uremia Program.
3. A. Chapiro and A. Jendrychowski-Bonamour, C. R. Acad. Sc. Paris, 270 (1970) 27-30.

BIOMATERIALS FOR DIRECT SKELETAL ATTACHMENT

P. H. Newell, Jr., T. A. Krouskop,
R. J. Jendrucko, and B.K. Chakraborty
Bioengineering Program
Texas A&M University
College Station, Texas 77843

ABSTRACT

Direct Skeletal Attachment of a prosthetic appliance currently has three major problems associated with it: they are 1) attachment to the bone, 2) intrusion through the skin, and 3) development of a prosthesis that is worthy of permanent attachment (activity to solve this problem is described in reference 1). The systematic development of biocompatible materials for use in direct skeletal attachment requires an understanding of the metabolic processes at the tissue interface. In this regard, a set of activities to develop a quantitative understanding of mechanical stress induced bone growth and to develop composite systems for use as artificial skin has been initiated. A conceptual model that relates the phenomenological stimuli responsible for bone remodeling has been developed as a preliminary exercise to quantify Wolff's Law. A set of thermodynamic and chemical kinetic equations have been formulated. Their use in describing the formulation of crystalline hydroxypatite is presented. A macroscopic structural analysis of the bone pylon interface has been conducted and the results of this investigation are discussed in relation to establishing pylon design criteria. Finally, the use of velour-α-amino acid composites and the use of scar tissue is discussed as a means of producing a stable intrusion through the skin.

INTRODUCTION

Since World War II, the technology associated with developing prosthetic systems has progressed at a rapid rate. This progress is due primarily to advances in polymer chemistry and increased emphasis on engineering analysis and design of systems which utilize the new materials to satisfy the criteria of total cosmesis. Accordingly, the prospect for rehabilitating today's amputee is much better than it was twenty years ago. However, the current generation of clinically useful artificial limbs have a major limitation associated with them; they rely on transmitting loads through the soft tissue which often results in tissue necrosis and subsequent changes in stump geometry and marginal control of the assistive unit.

Since this problem cannot be solved using conventional prosthetic tissue interface systems, the technique of direct skeletal attachment has recently become a major area of rehabilitation engineering research and development. The body of knowledge in this area has expanded quickly and a wide range of material systems comprising ceramics, metals, and polymeric composites have been developed to satisfy the biocompatibility requirements at the bone interface and the skin interface. But the development of these materials has been based on empirical studies for which no theoretical basis exists. The material characteristics which are necessary for stable interfacing cannot be clearly established until the basic mechanisms that control the metabolic response at the tissue-prosthesis interface are delineated. In this regard, the most unique characteristics of a living system is its ability to selectively concentrate chemical species against their total potential gradient in response to stimuli. This concentration occurs at a site corresponding to the stimulus and the process is known as active transport. At Texas A&M University, the Bioengineering Program has developed a working hypothesis to explain quantitatively this transport process. The hypothesis is that the potential for active transport exists when:

1. at least two phases are present in a system which are separated by a semi-permeable membrane which is permeable to at least one species.

2. reversible chemical reactions involving the permeable species occur in each phase, and

3. a source and sink for the activation energy of phosphorylation and synthesis reactions in living systems.

We have analytically and experimentally verified this hypothesis for molecular transport. It is the understanding of this coupling between the chemical kinetics and mass transport in living systems that renders an engineering analysis of metabolic modeling possible. The specific aims of this approach have been (a) to model the bone growth and remodeling phenomena in response to applied mechanical stress, (b) to design the geometry of the bone-biomaterial interface in order to optimize stress induced bone ingrowth, and (c) to develop material systems which can be used to provide a permanent skin-biomaterial interface.

MODELS OF BONE GROWTH

Bone behaves as a stimulus-response adaptive system and thus the quantification of Wolff's Law would be ideally represented by an isomorphous model which includes all the components and functions of the actual system. However modeling biological processes at the molecular level requires a detailed understanding of the component processes and the physical and chemical parameters that characterize the subsystems. Since current knowledge relating to the subsystems consists of open loop models and does not provide detailed quantitative information, a closed loop phenomenological model (Figure 1) has been developed to set priorities for further experimental programs and as a first step in understanding the behavior of the bone.

The sequence of events which comprises bone remodeling result from the net effect of the osteoblastic and osteoclastic activity. The signals that are critical to the initiation and control of cellular activity are not understood quantitatively. Based on current information (2,3), the control signals have been grouped into two categories: 1) those related to the mechanical effects, and (2) those pertaining to

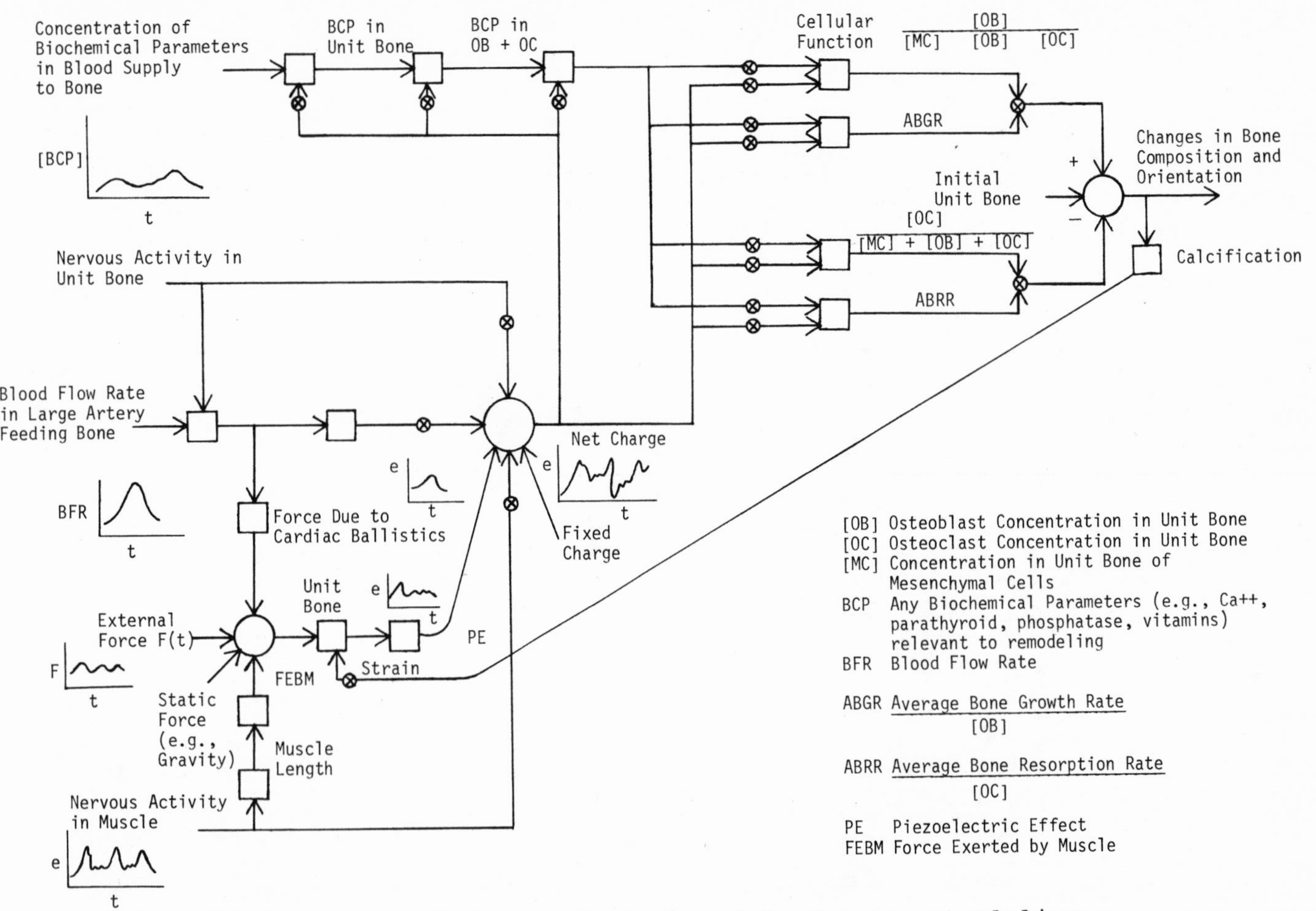

Figure 1. A phenomenological model of bone remodeling

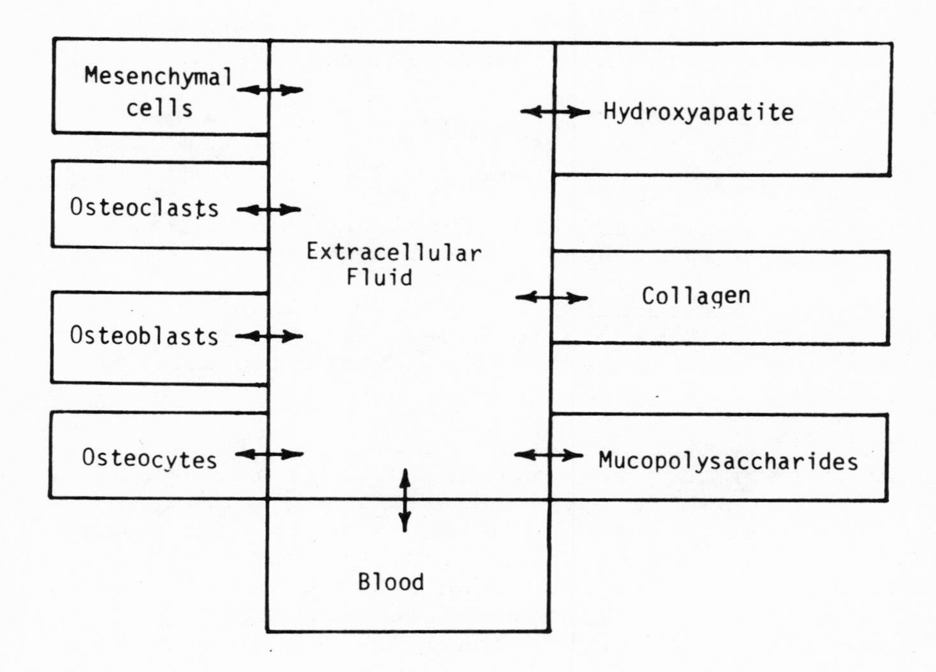

Figure 2. A compartmental representation of bone

cellular biochemistry. In the preliminary model, the dynamic changes in the local composition and orientation in bone are taken as the primary output variables.

The mechanical signals are likely to be transduced in the bony tissue to other forms of energy for stimulating osteoblastic and osteoclastic activities. In fact, it is becoming increasingly clear that the bony matrix transduces mechanical forces into electrical signals. The major sources for generating the electrical signals have been identified as:

1. mechanical forces (static and dynamic)

2. nervous activity (due to muscle action potentials and central nervous system impulses directed to the bone), and

3. blood flow effects (including streaming potential generation and cardiac ballistics).

The net bioelectric signal is the summation of these effects and the fixed charge which is characteristic of the unit bone when added in the spatial coordinates of the osteon.

The biochemical signals are presumed to be transmitted through the extracellular fluid that is in direct contact with the cells. The biochemical parameters, important in bone biodynamics, are numerous. However by using specialized transfer functions, the effects of various key components are being studied by selecting representatives of the important groups, such as calcium and phosphorus ions, parathyroid and calcitonin, acid and alkaline phosphatases, etc. The possible effects of the electrical signals in the unit bone on the biochemical parameters have also been considered as reflected in their effect on ionization states.

In a concurrent modeling activity, the bone has also been considered as a set of discrete phases in contact with one another (Figure 2). In this model the extracellular fluid is viewed as interfacing with the other phases such as collagen, hydroxyapatite mucopolysaccharides as well as the intact cells. The basis for this multicompartmental model approach lies in the assumption that materials exchanged between any two compartments can be described in terms of quasistatic processes occurring between a series of equilibrium state points.

For any two of the interfacing compartment, the Gibb's equation for an open system relates mechanical forces (f) imposed on the system to changes in 1) the system geometry (changes in length, ℓ, for a one dimensional change), 2) component masses (n, the number of moles) and 3) chemical potentials (μ).

As an example, consider a one component, two phase system. The Gibbs equation is

$$(dG)_{T,P} = Fd\ell + \mu dn.$$

For a change between any two equilibrium states

$$(\Delta G)_{T,P} = 0.$$

For the case where F and μ are constant,

$$F\Delta\ell + \mu\Delta n = 0.$$

Thus, the number of moles lost from the solid under one-dimensional compression resulting in a length change

$\Delta\ell$ is $\Delta n = \frac{F}{\mu}\Delta\ell.$

This treatment may be extended to liquid and solid phases in which chemical reaction occur. Consider the same two-phase system but with species A reacting reversibly to form compound B according to : $A \rightleftarrows B$. During the reversible compression of the solid and the coupled chemical reaction, the equilibrium requirements are:

$$0 = \int_{\ell_1}^{\ell_2} Fd\ell + \int_{n_1}^{n_2} \mu_A^{solid}\, dn$$ (Gibbs equation for solid)

$$\mu_A^{solid} = \mu_A^{solution}$$ (Equation of phase equilibrium)

$$\mu_A^{solution} = \mu_A^{o} + RT \ln \frac{n_A}{n_{total}}$$ (Definition of chemical potential in solution)

$$\mu_A^{solution} = \mu_B^{solution}$$ (Equation of chemical equilibrium)

Similarly for a two-phase system with many components, this methodology may be expanded to yield a generalized equation set:

$$0 = F d\ell + \Sigma \mu_i dn_i$$ (Gibbs equation)

$$\mu_i^{\alpha} = \mu_i^{\beta}$$ (Equations of phase equilibrium, α, β phases)

$$\mu_i = \mu_i^{o}(T,P) + RT\ell n \frac{n_i}{n_{total}}$$ (Solution chemical potentials)

$$\mu_A + \mu_B = \mu_C + \mu_D$$ (Equation of chemical equilibrium for a reaction of the type $A+B \rightleftharpoons C+D$)

In addition to these equations the following two constraints apply:

$$\ell dF + \Sigma n_i d\mu_i = 0$$ (Gibbs-Duhem)

$$dn_A^{\alpha} + dn_B^{\alpha} + dn_A^{\beta} + dn_B^{\beta} + --- = o$$ (total system mass balance)

It can be shown that, for a two-phase solid-liquid system with chemical reactions and F related to ℓ through the constitutive equiation $F = k\ell$, the number of equations which can be written is equal to three times the number of components plus the number of independent chemical equilibrium equations plus six. By a similar generalization it can be shown that the number of variables equal to four times the number of components plus four. Thus, this system is solvable only when the number of equations equals or exceeds the number of variables or when the number of components are less than or equal to two plus the number of independent chemical equilibrium equations.

The initial effort utilizing this approach has been concentrated on analyzing a system consisting of a solid hydroxyapatite phase and a liquid phase that simulates extracellular fluid. The equation for apatite formation has been idealized as:

$$10Ca^{++} + 2OH^{-} + 6PO_4 \rightleftarrows Ca_{10}(PO_4)_6(OH)_2$$

This analysis is being used to provide insight into the kinetics involved in the formation of hydroxyapatite. Upon completing this analysis, the results will be compared with the experimental results from other laboratories (3).

In order to complete the first iteration molecular model for apatite formation in a single osteon, a nonsteady state diffusion model has been formulated (Figure 3.) This model is based on a form of the diffusion equation that allows for variable diffusivity in space and time coordinates. The model is being used to predict the calcium-content time course of bone mineralization. When coupled with the thermodynamic model for crystal formation and growth, this formulation should adequately describe the calcification function described in the phenomenological model.

DESIGN CRITERIA FOR INTERFACING BONE PYLON

While the role of stress in tissue ingrowth will be better understood through the modeling of the pressure-induced bone growth phenomenon, how desirable stress patterns can be developed at the bone-biomaterial interface is a separate problem that must be resolved. Currently, there are no data available regarding the interrelationship between the geometry of a bone pylon and the stresses at the bone-biomaterial interface. Therefore, a finite element model of the bone-pylon system has been developed and is being used to analyze the system and to generate quantitative guidelines for the design of bone pylons.

In order to make the mathematical model react in a manner that closely resembles the lower extremity, a set of support conditions were derived to approximate the hip. In conjunction with these conditions data about muscular forces have been used to provide a set of boundary conditions that simulate the _in-vivo_ conditions experienced by a bone-pylon system.

$$\frac{\partial c}{\partial t} = \bar{\nabla} . D_E \bar{\nabla} c - R$$

c = concentration of any species

$\bar{\nabla}$ = vector operator

R = rate of hydroxyapatite deposition; assumed $R = f(C_{Ca^{++}})$

$$D_E = \frac{\varepsilon}{K_T^2} . D$$

ε = open void fraction

K_T = f (average diffusion path length)

D = diffusivities of species in a dilute aqueous solution

Figure 3. A model for diffusion in bone

A no slip condition has been imposed at the interface between the pylon and the bone surface to evaluate the magnitude of the force that must be developed at the interface to prevent slippage. The results of this analysis have been reduced and used to prepare a set of iso-stress contour plots that depict the effect of pylon shape on the loading experienced by the bone (Figure 4).

In the future, this model will be revised to include the thermodynamic-chemical kinetic model of bone growth. By using this tool, it will be possible to study, in detail, the time course of the stress patterns and to design a pylon system that utilizes a resorbably medullary pin for initial stabilization but relies exclusively on cortical end attachment for long term strength without having to wait for the results of experimental animal studies.

DESIGN CRITERIA FOR THE SKIN BIOMATERIAL INTERFACE

The skin interface problem is an active area of research in many laboratories (4-10) and this work has resulted in a multitude of experimental designs which have met with varying degrees of success.

In this program three possible approaches to skin-interfacing at the site of direct skeletal attachment are currently being considered (Figure 5). Based on the results of Hall's work (5) and that of Miller and Brooks (6), an evaluation of the "Collar Stud" method is being conducted to evaluate various biomaterials, design and construction details, and procedures for surgical insertion. The modification of a double-walled velour pouch technique, reported for electrode implantation by McCutcheon, et. al. (8) is also being explored as a means of producing a viable intrusion through the skin.

These three schemes have been developed in an attempt to solve the problem associated with bonding to the dermal layers the sloughing off of the artificial skin as a result of nomal skin growth. The material system, a composite of textile velour and poly-a-amino acid (DL-leucine-methionine) are being used in these experiments since this type of composite skin substitute has been reported to possess desirable permeability characteristics and microbe sealing properties (7).

Figure 4. A typical bone-pylon model and stress plot, in this case a maximum principal stress plot with marrow cavity full.

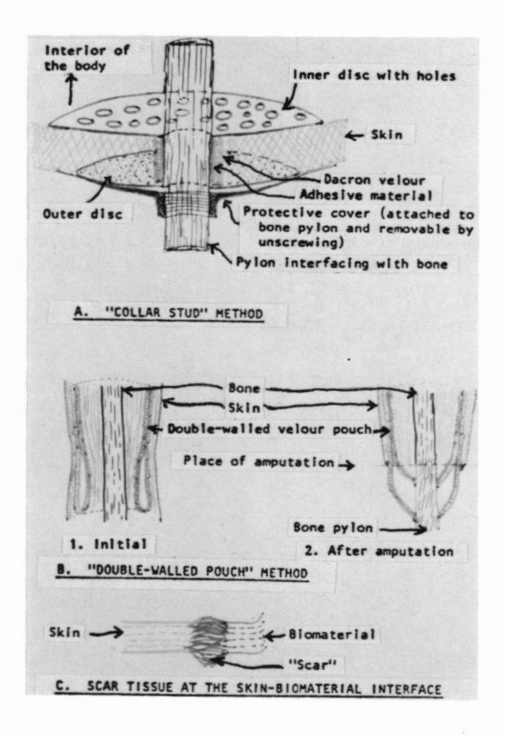

Figure 5. Alternative methods to achieve intrusion through the skin.

Finally, a third approach for developing an effective interface between skin and a prosthetic implant has been to explore the potential of using scar tissue. Scar tissue has a morphology grossly different from that of skin, however there is a need for more quantitative data on fibrous tissue fromation, the nature of its metabolism and its biomechanical properties before the potential of this scheme can be fully evaluated.

SUMMARY

The interim results presented in this paper relate to developing a theoretical base for the design of materials which can be used for direct skeletal attachment of prosthetic equipment. In this regard, a closed-loop phenomenological model of bone dynamics has been formulated and used to develop a first iteration model which describes the chemical kinetics, thermodynamics, and diffusion of bone mineralization as a function of the mechanical loading on the system. A mechanical model of a bone-pylon system has been developed to study the effects of pylon shape on inducing bone ingrowth; and finally a series of skin interfacing techniques are described. By continuing this research effort in conjunction with a program to develop a modular endoskeletal prosthetic system, it is envisioned that direct skeletal attachment will become a clinically accepted procedure within the next ten years.

ACKNOWLEDGEMENTS

The authors gratefully acknowledge support of the research described in this paper from the Social and Rehabilitation Service by means of Grant No. 23-P-55823/6.

REFERENCES

1. Newell, P. H., Jr., T. A. Krouskop, and L. A. Leavitt, "A Lightweight Modular Endoskeletal Hip-Disarticulation Prosthesis," presented at The International Society for Prosthetics and Orthotics Symposium, August 20-24, 1972, Sydney, New South Wales.
2. Bassett, C. A. L., 1971. "Biophysical Principles

Affecting Bone Structure, in the Biochemistry and Physiology of Bone," G. H. Bourne (Ed.), Vol. 3, p. 1-76.

3. Justus, R., J. H. Luft, 1970. "A Mechanochemical Hypothesis for Bone Remodeling induced by Mechanical Stress." Calcified Tissue Res. 5:222-235.
4. Hulbert, S. F., F. Y. Young, R. S. Mathews, J. J. Klawitter, C. D. Talbert, F. H. Stelling, 1970. "Potential of Ceramic Materials as Permanentl Implantable Skeletal Prosthesis," J. Biomedical Materials Res. 4:433-456.
5. Hall, C. V., D. Liotta, M. E. DeBakey, 1966. "Artificial Skin," Trans. Amer. Society Artificial Internal Organs. 12:340-342.
6. Miller, J. and C. E. Brooks, 1971. "Problems related to Maintenance of Chronic Percutaneous Electronic Leads," J. Biomedical Materials Res. Symp. Vol. 2 (pt. 1):251-257.
7. Skornik, W. A., D. P. Dressler, J. H. Richard, 1962. "Adherence of Prosthetic Skin," J. Biomedical Materials Res., 2:447-456.
8. McCutcheon, J., M. Evans, R. F. Wibel, 1973. "A Fabric Pouch for Maintenance of Multiple Chronically Implanted Lead Terminations," 4th Annual Meeting of Biomedical Engineering Society.
9. Vasko, K.A., and R.O. Rawson, 1967. A chronic nonreactive percutaneous lead system. Trans. Amer. Society Artificial Internal Organs, 13: 143-145.
10. Hegyeh, R.J., 1971. The Physiological and Biomechanical Basis for Soft Tissue implants. Biomaterials, A.L. Bement, Ed., p. 207-233.

THE ROLE OF RAPID POLYMERIZING ACRYLIC DISPERSIONS IN DENTAL PROSTHESES

F. F. Koblitz, S. D. Steen and J. F. Glenn

Central Research Laboratory
Dentsply International, Inc.
York, Pennsylvania 17404

INTRODUCTION

Dental prostheses, particularly dentures, are very important biomedical devices. More than thirty-five million people in the U. S. depend on dentures for an attractive appearance and the ability to speak and eat naturally. Dentures are constructed predominantly from acrylic polymers and artificial porcelain or acrylic teeth. Consequently, the clinical and practical utility of dentures and the oral health of a large segment of our population depends on the properties of acrylic biopolymers.

The American Dental Association has outlined the requirements and specifications for artifical teeth and denture base polymers (3). These requirements have been adequate in the past for the development of conventional acrylic dough type denture base polymers, but they were inadequate for the characterization and evaluation of the acceptability of rapid polymerizing acrylic dispersions as denture base constituents. It was necessary to study the acrylic dispersions in terms of volume fraction and particle size distribution of the dispersed phase, viscosity behavior, density of the dispersions before and after polymerization, adhesion to artificial teeth, and microstructure.

EXPERIMENTAL

A. Materials

Two suspension polymerized poly(methyl methacrylate) bead homopolymers with molecular weights of 160,000 (Polymer A) and 950,000 (Polymer B) were evaluated singly and in a 50/50 weight blend (Polymer C) as the dispersed phases of model dispersions.

The dispersed phase of the dispersions contained, by weight: 97.97% polymer beads, 0.9% benzoyl peroxide (98% active), 1% red acetate fibers, 0.03% red pigments, and 0.1% TiO_2 pigment.

The liquid phase of the dispersions contained: 84% methylmethacrylate, 5% ethylenedimethacrylate, 0.5% distilled dimethyl-p-toluidine, 10% methyl-methacrylate-ethylmethacrylate copolymer, and 0.5% ultraviolet absorber (benzophenone derivative).

B. Test Methods

American Dental Association Specification No. 12 (3) was used to test the properties of experimental dispersions and relate them to the properties of the conventional denture base polymer which meets the requirement of the dental profession. (4)

Polarized light microscopy and scanning electron microscopy were used to determine the internal morphology of the dispersions before and after polymerization.

True powder densities and the density of the polymerized dispersions were determined using a pycnometer by standard methods. (5) The density of the liquid dispersions was determined by hydrometer. (5)

The overall volumetric shrinkage of the dispersions which could be expected during denture manufacture was calculated from the densities of the dispersions determined immediately before polymerization and from the densities of the set masses one day after polymerization.

Particle sizes were determined by Coulter Counter using a 400 micron orifice.

Dispersion viscosities were determined using a Brookfield Viscometer, model RVT.

Figures 1 to 4 show a procedure used by dental laboratories (6) to construct dentures from acrylic dispersions. A wax model denture mounted on a dental plaster model of the patient's mouth is initially mounted in an articulator (a device which simulates the relationships of the upper and lower jaws). The proper functioning of the wax denture is confirmed, and an index of the positions of the teeth is made using dental plaster. The wax model denture is then encased in a rigid mold, and the wax is melted by immersion in boiling water. The artificial teeth are retained in position in the mold. The surfaces of the model of the mouth and the mold are carefully cleaned and release film is applied to both. The acrylic dispersion is poured into the mold and allowed to deaerate. It is then polymerized in a pressure cooker for 30 minutes at 60°C. under 20 psi air pressure. This method was used to prepare acrylic dispersion dentures for subsequent clinical testing and to prepare adhesion test specimens.

The adhesion of the denture base polymer to artificial plastic teeth was determined using an Instron Universal Testing Machine. Artificial plastic teeth were embedded in the denture base polymer by the denture processing methods (6) described above. The artificial teeth were embedded on their lingual (inside) surface parallel to the plane of the specimen with the long axis of the teeth perpendicular to the long axis of the specimen bar. The test bars were clamped in a jig perpendicular to the direction of stress in the Instron jaws. Stress was imposed perpendicular to the surface of the specimen to break the teeth upward from the embedding interface in a cantilever mode. The rate of pull was 0.1 in/min.

RESULTS AND DISCUSSION

The retention of the spherical shape of acrylic beads after polymerization of a self-curing acrylic bead-monomer mixture has been reported. (7) This report was confirmed by polarized light microscopy on a polymerized model acrylic dispersion (Fig. 5). Retention of particle shape and relative particle size distribution were observed. Refractive index differences between matrix and beads and birefringence

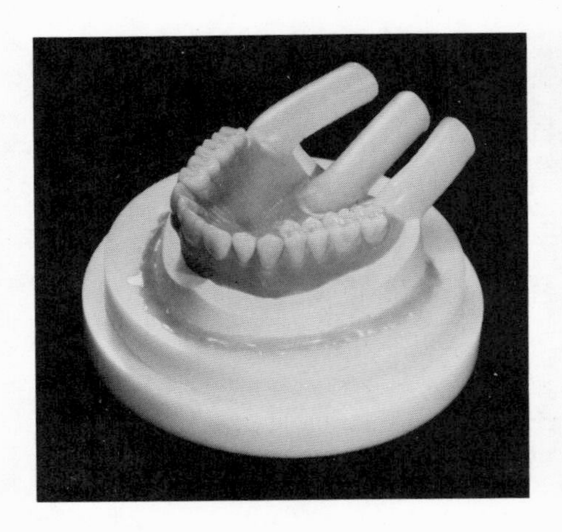

Fig. 1. Wax model denture prepared for processing

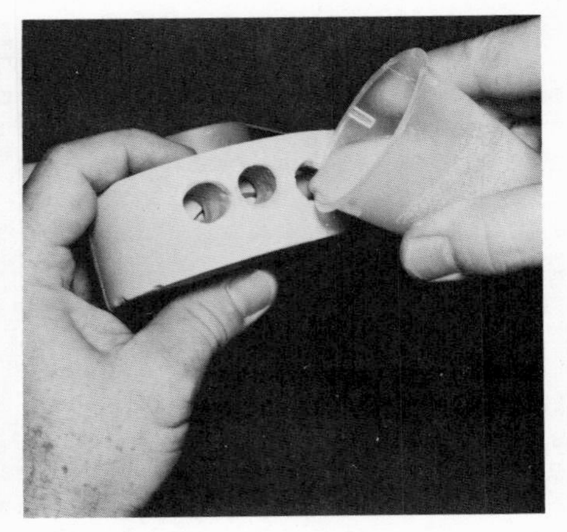

Fig. 2. Casting an Acrylic Dispersion into a Denture Mold

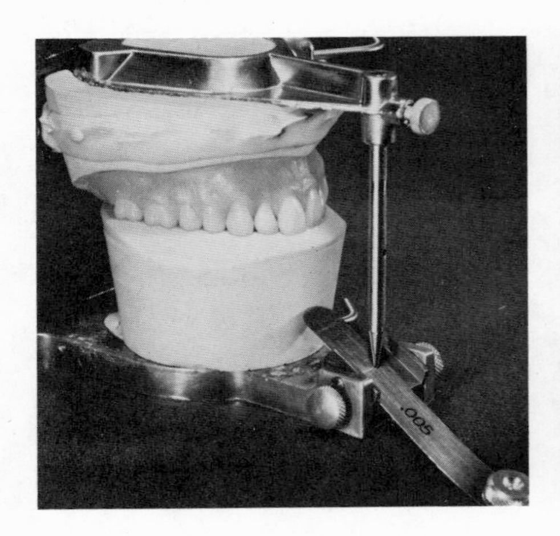

Fig. 3. Denture Mounted in an Articulator with Occlusal Index

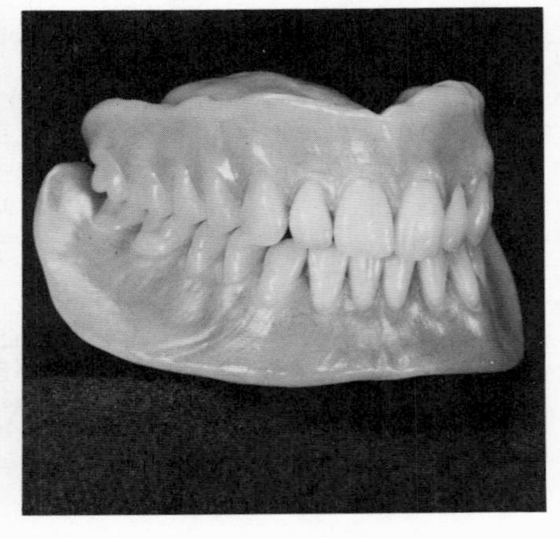

Fig. 4. Complete Dentures Constructed from Acrylic Dispersion Denture Base

in fibers and strained beads were also evident. Fracture studies by polarized light and scanning electron microscopy also indicated the retention of spherical bead shape and showed the packing behavior (uniformity of dispersion) of the beads (Fig. 6,7). Fracture and de-wetting at bead matrix interfaces indicated a critical volume of matrix was required for bead wetting and dispersing. Insufficient liquid matrix inhibited mixing of the dispersion resulting in incomplete wetting of beads and reduced mechanical properties (Fig. 8 and Table I).

TABLE I

Effect of Volume Fraction of Dispersed Phase on Mechanical Properties (Model Dispersion)

	63%	59%	56%
Impact Strength (Unnotched-ft-lb/in.)	1.8	2.1	2.2
Diametral Tensile Strength (psi)	4650	5350	4900
Transverse Deflection (mm) (not immersed)			
3500 g load	1.7	1.8	1.8
5000 g load	3.4	3.6	3.4
breaking load	5500	6500	6000

Conversely, a volume fraction of liquid near the optimum resulted in closely packed spherical beads with no de-wetting at interfaces. Fig. 9 shows the fracture surface of an artifical acrylic tooth-acrylic dispersion denture. The uniformly "pebbled" denture base surface "telegraphed" the homogeneous dispersion of the acrylic beads. Also evident was the high degree of fracture resistance of the tooth-dispersion interface. The characteristics of a model acrylic dispersion and a conventional denture resin are shown in Table II.

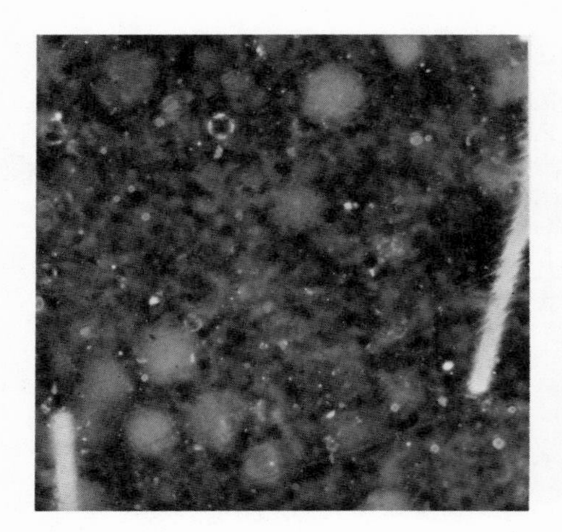

Fig. 5. Internal Morphology of Polymerized Fibered Acrylic Dispersion. Polarized light Microscopy (13X) Blue Filter.

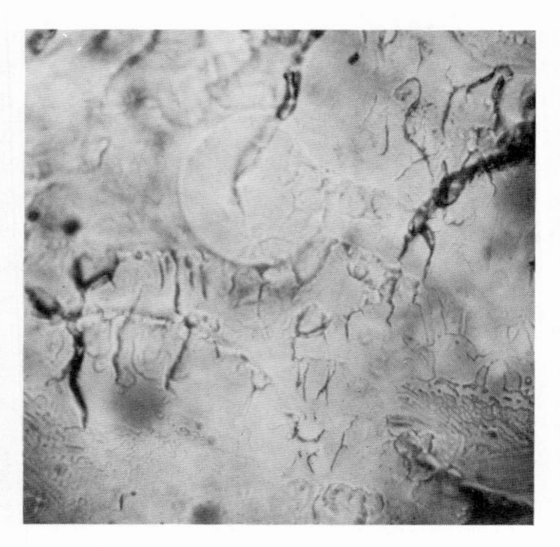

Fig. 6. Fracture Surface of Polymerized Acrylic Dispersion. Bead Volume Fraction = 67%. Polarized Microscopy (80X).

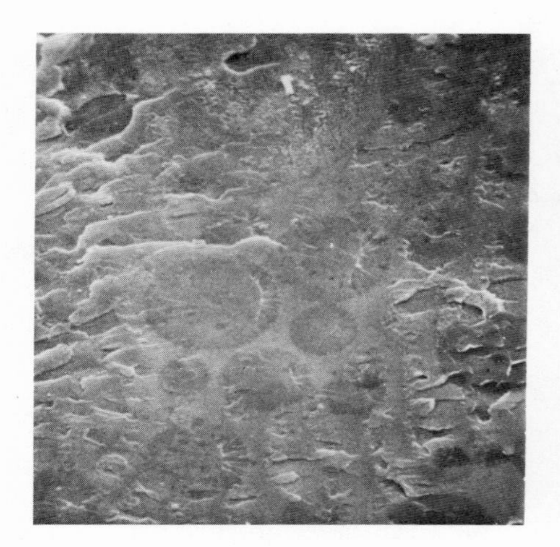

Fig. 7. Fracture Surface of Polymerized Acrylic Dispersion (60% v. beads). SEM 300X.

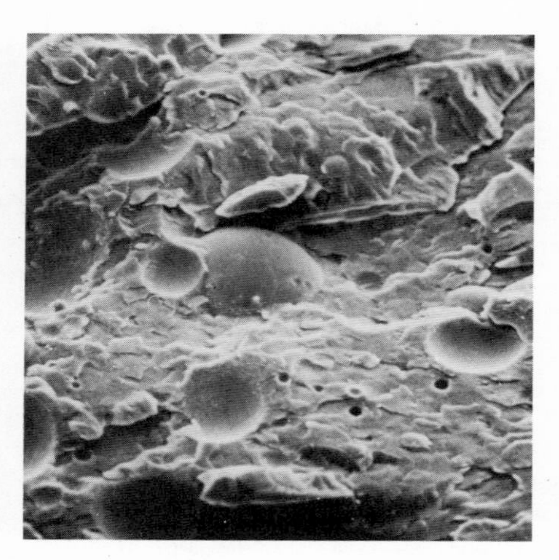

Fig. 8. Fracture Surface of Polymerized Acrylic Dispersion (68% v. beads). SEM 300X.

TABLE II

Properties of Polymerized Denture Bases

Property	Model Dispersion (Polymer C)	Conventional Denture Base (1)
Toxicity	no toxic effects or mucosal irritation	no toxic effects or mucosal irritation
Surface Characteristics	non-porous, no defects, high polish, uniform color	non-porous, no defects, high polish, uniform color
Color stability	v.sl. fading	v.sl. fading
Water sorption (mg/cm^2)	0.6	0.6
Water extraction (mg/cm^2)	0.02	0.02
Transverse Deflection (mm) 3500 g load	1.9	1.8
5000 g load	4.4	4.3
Shore D Hardness	86	87
Impact Strength (ft-lb unnotched)	2.2	3.1
Diametral Tensile strength (psi)	5350	5550
Density (g/cc at 23°C)	1.17	1.18
Adhesion to Plastic Teeth (2) (lb.)	53 (cohesive in teeth)	54.4 (cohesive in teeth)

(1) Characterized Lucitone (L. D. Caulk Co.), a compression molded polymer, cured at 180°F., not directly comparable to acrylic dispersions but representative of best quality dental polymers.

(2) Biotone plastic teeth (Dentsply International).

The bond strength of polymerized dispersion to artificial teeth exceeded the cohesive strength of the teeth. No fractures were observed at the interface.

These observations prompted a review of volume fraction effects, (8,9) packing behavior, (8,9,10) and viscosity behavior (8) of similar dispersion systems and further investigations of these subjects with rapid polymerizing denture base dispersions.

Particle size distributions of acrylic beads were studied by Coulter Counter (TABLE III) and by dispersion in silicone oil (Fig. 10).

TABLE III

Particle Size Distribution

(% by weight in each fraction)

Diameter Range of each fraction in microns	Polymer A	Polymer B	50/50 Blend A/B
0-11	7.9	13.0	6.9
11-32	4.6	14.4	7.2
32-64	4.3	54.0	28.2
64-100	9.6	18.0	9.5
100-128	5.1	-	4.7
128-160	15.4	-	14.2
160-200	53.1	-	29.3
	100.0	99.4	100.0

The requirements for a polydisperse system were indicated qualitatively, although particle swelling and dissolution in the liquid matrix precluded the calculation of an optimum particle size distribution.

In practice, the dispersions exhibited substantial viscosity rises approximately 4 minutes after mixing (Table IV).

Light and scanning electron micrographs failed to clarify the relative contributions of dissolution of beads from the dispersed phase and sorption of methyl methacrylate monomer from the liquid phase. The dissolution of the beads into the liquid phase prior and during polymerization would tend to increase the average molecular weight of the matrix and improve the overall mechanical properties of the polymerized dispersion.

Although the microstructure studies failed to clarify the extent of swelling of beads and dissolution of beads, they did indicate that polydispersity of the

TABLE IV

Dispersion Viscosities (1)
(in centipoises)

Polymer	1 min. 30 sec.	4 min.	5 min.
A	190	580	850
B	1350	5400	9000
50/50 A/B	500	1200	1800
Dispersion A(2)	250	750	1400

(1) Brookfield RVT, #4 spindle, 100 rpm. Data in centipoises for a 59% v. dispersion of polymer in model dispersion liquid at 75°F.
(2) Duraflow Porit (Myerson Tooth Corp.), a commercial acrylic dispersion denture base.

dispersed bead phase was required for close dense random packing.

The results of viscosity studies shown in Table IV further indicated the desirability of a poly-disperse particle size distribution.

In practice, the critical bead volume fraction of a polydisperse system was approximately 65%. Above 65% by volume of dispersed phase the dispersions were too viscous and thixotropic to be successfully cast in denture molds. Bead volume fractions above about 65% caused incomplete wetting and mixing. A fracture surface of polymerized acrylic with a bead volume fraction of 67% is shown in Fig. 8. Dewetting occurred at bead-matrix interfaces during fracture. Thus, the approximate 65% critical bead volume appears to hold both for handling properties and for strength properties in the set mass. Figure 11 shows the strong adhesion and absence of dewetting observed for an acrylic dispersion with 60% by volume of beads bonded to an artificial tooth.

Observation of polymerized dispersions formulated with low volume fractions of beads indicated that their shrinkages would be sufficient to cause dimensional inaccuracies if they were used as materials of construction for dentures. Monodisperse bead distributions required substantially more liquid phase for their mixing and dispersal, leading to lower bead volume fractions and greater shrinkage

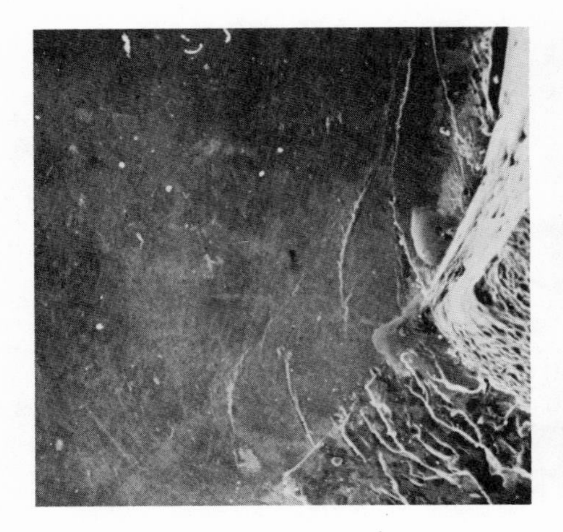

Fig. 9. Fracture Surface and Exterior Surface of Artificial Tooth (on left). Bonded to Polymerized Acrylic Bead Dispersion. SEM (50X)

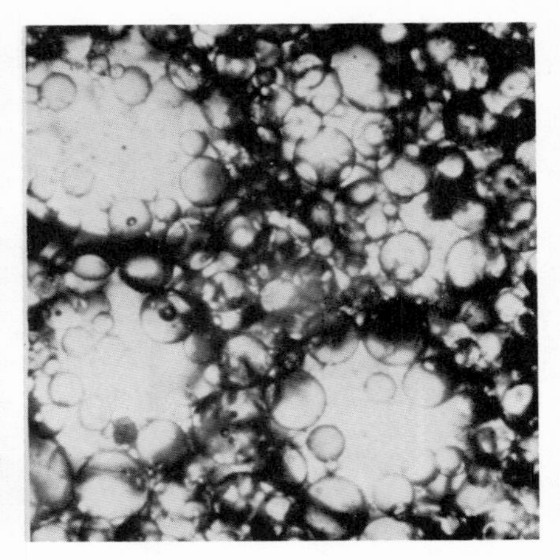

Fig. 10. Acrylic Beads Dispersed in Silicone Oil Polarized Light Microscopy (40X)

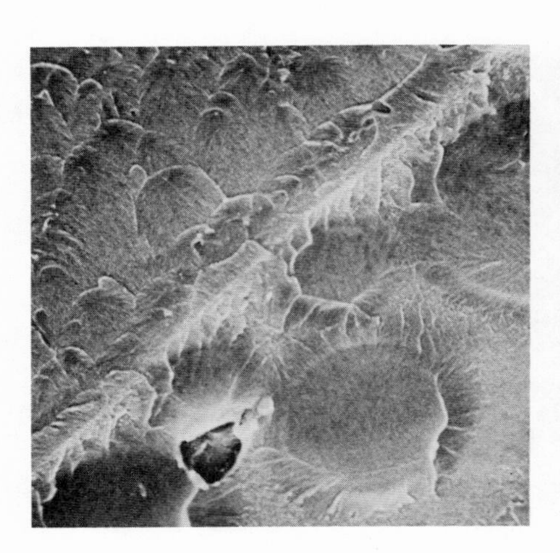

Fig. 11. Fracture Surface of Artificial Tooth Bonded to Polymerized Acrylic Bead Dispersion (60% v. beads) SEM (1000X)

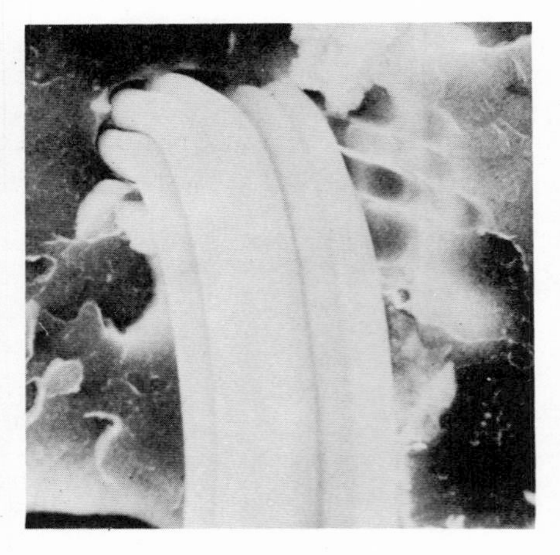

Fig. 12. Fibers Protruding from Fracture Surface of Denture Base (Pour-N-Cure, Coe Laboratories) (55% v. beads).

during polymerization. Fig. 12 shows a polymerized dispersion with 55% by volume of the monodisperse beads. A fiber strand is protruding from the fracture surface. Fibers are customarily added to characterize the otherwise featureless and highly glossy surface of dentures. In this polymerized dispersion (Dispersion B), shrinkage caused a large enough gap to allow ingress of mouth fluids and consequent bacterial action if it were to be used in the mouth.

Further blending studies on polymers and the inclusion of a polymer in the liquid phase of dispersions reduced the total volumetric processing shrinkage of the dispersions as shown in TABLE V. Dentures constructed from dispersions with shrinkages in the range of 5% volumetric shrinkage were found to be clinically acceptable by occlusal index and vertical height determinations in articulators as shown in Fig. 13. The low shrinkage dispersions met the requirements of American Dental Association Specification #12 for denture base polymer (TABLE II). Oral mucosal irritation and toxicity tests (11) demonstrated the biocompatibility of the cured acrylic dispersions.

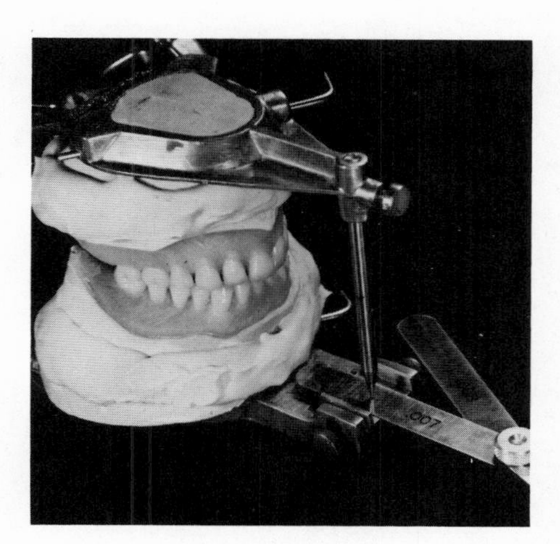

Fig. 13. Complete dentures processed from model dispersion mounted in articulator

TABLE V

Densities (1) and Processing Shrinkages (2) of Acrylic Powder-Liquid Mixtures

Mixture	Bead Density	Liquid Density	Unpolymerized Density Found	Unpolymerized Density Calculated	Polymerized Density (5)
Polymer B (59% v. beads)	1.19	.97	1.09	1.10	1.17(6.0)
Polymer C (59% b. beads)	1.20	.97	1.10	1.11	1.16(5.2)
Dispersion A (59% v. beads)	1.22	.95	1.09	1.11	1.17(6.8)
Conventional Denture Base	1.21	.94	1.08	1.12	1.18(5.1) (4)

(1) Densities in g/cc at 23°C.
(2) Processing shrinkage in % decrease in volume of dispersion calculated from increase in density during processing.
(3) Unpolymerized density calculated from volume fractions and densities of beads and liquids.
(4) A calculated value from an unpolymerized density of 1.12. The "unpolymerized" dough contained air which contributed to a low density of 1.08.
(5) Shrinkage during processing shown in ( ).

CONCLUSIONS

The test methods of American Dental Association Specification #12, standard methods for measuring viscosities and densities, polarized light microscopy, scanning electron microscopy, oral toxicity, and mucosal irritation tests were used to demonstrate that rapid polymerizing dispersions can meet the standards of the dental profession as constitutents of dental prostheses. The acrylic dispersions consisted of spherical acrylic beads stirred into rapidly polymerizing matrices. The dispersed beads retained their spherical shape in the polymerized matrices.

Critical volume fraction phenomena were observed for the handling properties and microstructure of the dispersions. The critical bead volume fraction for handling and uniform microstructure correlated closely with the critical bead volume fraction required for dimensional accuracy and mechanical properties of biopolymers designed for use as denture bases.

REFERENCES

1. Marketing Committee, American Dental Trade Association, The Professional Market, 1970, A Survey Report.
2. B. Duane Moen and W. E. Poetsch, Bureau of Economic Research and Statistics, American Dental Association, JADA, 81, 25-36 (July, 1970).
3. American Dental Association, Guide to Dental Materials and Devices, 96-105, 201-206, Ed. 6 (1972-1973).
4. G. C. Paffenbarger and N. W. Rupp. Research techniques used in evaluating dental materials. JADA, 86, 643-651 (March, 1973).
5. Arnold Weissberger, Physical Methods of Organic Chemistry, vol. 1, 148-180, Ed. 3 (1959).
6. Dentsply International, Inc., Suggested Laboratory Techniques for the TruPour Fluid Denture Resin System, (1973).
7. D. C. Smith, Brit Dent. J. 111:9 (1961).
8. D. O. Lee, J. Paint Tech. 42, no. 550, 579-87 (1970).
9. G. P. Bierwagen, J. Paint Tech. 44, no. 574, 45-55 (1972).
10. R. H. Beresford, Nature, 224, 550-3 (1969).
11. Dr. R. Schlesinger, Bio-Technics Laboratories, private communication.

ADHESION MECHANISM FOR POLYURETHANE PREPOLYMERS BONDING BIOLOGICAL TISSUE

P. Y. Wang

Institute of Biomedical Engineering

University of Toronto

Toronto M5S 1A4, Canada

INTRODUCTION

A polyurethane prepolymer was first applied to living biological tissue in 1958. Unfortunately, subsequent clinical evaluation of this adhesive in an extremely demanding service condition as internal fixation for bone fractures was initiated before adequate scientific studies were made. The results were generally very disappointing (1). Later, another polyurethane prepolymer preparation was used satisfactorily as a coating to supplement sutures in vascular surgery (2). Recently, various polyurethane prepolymer systems have been evaluated as linings for dental restoration materials (3). The principle factors which control the adhesion of this type of prepolymer system to living biological tissues are not well established. There are reports indicating that pyridine is necessary for adhesion of the prepolymer to isolated beef muscle tissue; while other work has shown that excess amount of this amine is deterimental to bonding (4). It has also been noted on several occasions that an excess of the diisocyanate in the prepolymer mixture is necessary for good adhesion to the biological tissue (5,6). In previous publications (4,5), the preparation and some properties of fast setting polyurethane prepolymer systems have been described. With the adaption of a method to compare the bonding *in vivo*, (5) the effect of various components of the fast setting prepolymer system on adhesion is now evaluated.

MATERIAL AND METHODS

The preparation of the fast setting polyurethane prepolymer system has been described in detail elsewhere. (5) A sample composition of the prepolymer system used in the present study is given in Table 1. The change in bonding of the prepolymer systems as a function of their composition was evaluated on a pneumatic tensiometer using the dorsal skin of anesthetized Wistar rats as the substrate (5). The adhesion is reported as the arithmetic average of the individual pressure required to disrupt each of the four wounds on the same animal. The numerical values are the readings registered on the pressure gauge of the pneumatic device in pounds per square inch (psig), and NOT the tensile strength of the wound per unit area. All the experimental animals are sacrificed immediately after each test with chloroform overdose. The regulations made under the Animals for Research Act of Ontario (1968-1969) were observed during this project.

RESULTS

Adhesion of the Reactive Prepolymer

Samples of the prepolymer containing 1 millimole of castol oil and 2, 4, 6, and 12 millimole of tetrafluoro-1,3-phenylene diisocyanate, respectively, were prepared. These prepolymer samples also contained other minor additives in the amount shown in Table 1, but no tolylene diisocyanate. Bonding was evaluated on the skin incisions of the anesthetized animal. The results confirmed previous observation in this project (4,5). The adhesion effected by these samples of prepolymer on soft tissue was negligible (Table 2).

Effect of Tolylene Diisocyanate Concentration on Adhesion

Tolylene diisocyanate has been established to be the component essential for adhesion of the prepolymer prepared from fluorinated aromatic diisocyanate (4,5).

TABLE 1

COMPONENTS OF THE PREPOLYMER ADHESIVE

	mM/l
Castor Oil	1
Tetrafluoro-1,3-phenylene Diisocyanate	4.2
Tolylene Diisocyanate	4.0
Tetrahydrofuran	4
Pyridine	1
Surfactant	0.4% (W/W)

TABLE 2

EFFECT OF TETRAFLUORO-1,3-PHENYLENE DIISOCYANATE CONCENTRATION ON ADHESION

Tetrafluoro-1,3-Phenylene Diisocyanate (mM)	Adhesion* (psig)
2.0	3.2
4.0	3.7
6.0	3.8
8.0	3.6
12.0	2.7

*On the same scale, the strength of a 9-day wound healed naturally is 10.0.

Changes of adhesion to the soft tissue with the concentration of tolylene diisocyanate was investigated. The adhesion was found to increase to a maximum when the concentration of tolylene diisocyanate was about 4 millimole, and thereafter, it decreased slightly (Table 3). The decreasing value was due to gradual cohesive failure of the polyurethane material while it still adhered firmly to tissue. The variation of a measurable parameter such as adhesion with concentration suggests the occurrence of a chemical reaction at the polymer, tissue interface.

Effect of Pyridine Concentration on Adhesion

For a systematic evaluation on the role of pyridine in the adhesion process, the concentration of this amine in the prepolymer preparation was increased gradually. The adhesion was found to decrease linearly with increasing amount of pyridine (Table 4). Organic tertiary amines are known to promote crosslinking among reactive groups on the prepolymer structure during setting. This observation illustrates the influence of competing reactions and other chemical events in the over-all adhesion process. The excessive development of one reaction may result in poor adhesion. As later experiment indicates that the addition of tertiary amine is not the only means for further adjustment of the setting time of the prepolymer system.

Effect of Tetrahydrofuran Concentration on Adhesion

The inclusion of water miscible organic solvent such as tetrahydrofuran in the formulation as shown in Table 1 was to aid dispersion of the hydrophobic adhesive on the moist tissue surface, and to improve the general handling charactertistics of the adhesive. The amount added for this purpose prior to the systematic adhesion study was entirely arbitrary. The presence of such a small amount of tetrahydrofuran was not expected to influence adhesion. However, bonding study revealed other information due to its presence in the prepolymer system. For example, when approximately equal volume of the diluted prepolymer adhesive was applied on the tissue, the bonding was found to decrease with dilution (Table 5). The decrease in adhesion indicated that sufficient adhesive must be applied on the surface for adequate adhesion. However,

TABLE 3

EFFECT OF TOLYLENE DIISOCYANATE CONCENTRATION ON ADHESION

Tolylene Diisocyanate (mM)	Adhesion (psig)
0.0	3.7
0.3	5.7
2.0	8.7
4.0	9.9
6.0	9.9
8.0	8.2
12.0	7.0
15.5	7.5

TABLE 4

EFFECT OF PYRIDINE CONCENTRATION ON ADHESION

Pyridine (mM)	Adhesion (psig)
0.0	11.4
0.5	10.5
1.0	9.9
2.0	8.0
4.0	5.5

when adequate amount of the diluted adhesive was applied on the tissue surfaces until the net solid content on the exposed area was about the same after evaporation of the solvent, the adhesion was found to increase with dilution (Table 5). Thus, the dilution afforded better dispersion of the prepolymer in an aqueous environment as originally intended. The solvent also displaced the water on the applied area of the tissue surface, and facilitated disperion of these water molecules to the interior of the adhesive mass where they reacted with the isocyanate groups on the prepolymer. The displacement of water seemed to have liberated an increased amount of the available groups on the tissue surface for reaction at the interface with tolylene diisocyanate in the adhesive. The result was more efficient interfacial coupling, which was reflected as better adhesion on the tissue surface by the solidified polyurethane polymer.

The facilitated penetration of water into the adhesive mass also initiated foaming almost instantaneously without pyridine. At high dilution of the prepolymer by tetrahydrofuran, water-initiated curing became extremely rapid. The adhesive was often set before the tissue surfaces could be bonded.

Chemical Coupling Reaction on Soft Tissue

The chemical reactivity of tetrafluoro-1,3-phenylene diisocyanate is much higher than that of tolylene diisocyanate (7). It is unlikely that the latter may compete successfully in the formation of the prepolymer. The adjacent electron-releasing methyl group is also known to reduce the reactivity of the ortho-isocyanate function as compared to the paraisocyanate group in tolylene diisocyanate (8). Polyfunctional aromatic isocyanate is, therefore, not necessary for promoting adhesion of the prepolymer on soft tissue. Phenyl isocyanate, accordingly, should be able to promote adhesion similar to tolylene diisocyanate. This inference has been substantiated (Table 6). Hexyl isocyanate has also been found to be effective in promoting adhesion.

DISCUSSION

It is obvious from the experimental evidence presented that the adhesion of polyurethane prepolymers

TABLE 5

EFFECT OF TETRAHYDROFURAN CONCENTRATION ON ADHESION

Tetrahydrofuran (mM)	Adhesion (psig)	
	Equal volume of diluted prepolymer applied	Equal solid content of prepolymer applied
0.0	9.4	9.4
0.4		9.8
2.0	8.2	9.3
4.0	7.2	10.9
4.7		11.3
5.3		12.9
6.7		13.1

TABLE 6

TISSUE-POLYMER INTERFACE COUPLING AGENTS

	Adhesion (psig)
Tolylene Diisocyanate	10.0
Phenyl Isocyanate	8.7
Hexyl Isocyanate	10.6

on soft tissue is the result of two events at the tissue-polymer interface. The hydrophilic groups on the tissue surface must be converted by condensation with a compound, such as phenyl or hexyl isocyanate, into groups compatible with the prepolymer adhesive before its solidification. Contrary to the usual concept, (9) the results also demonstrate that adequate adhesion can be effected between tissue and synthetic polymers without direct covalent linkage. The most important step is the formation of a chemisorbed monolayer, probably discontinuous in nature, yet sufficient to permit adequate microscopic wetting of the tissue surface by the hydrophobic prepolymer. It is interesting to note that, the change in the surface energy of the tissue due to this chemisorbed monolayer is not immediately obscured by body fluids and its proteineous components. If rapid re-adsorption of water and proteins occured on the new surface, wetting and adhesion on the tissue by the polyurethane prepolymer would be impossible. The present results also lend some support to other studies on the structure of biological surface (10) which have indicated that the retention of hydrophobic cholesterol and fatty materials on arterial endothelium is preceded by changes of the vascular wall.

The chemisorption has assisted the adhesion of the polyurethane polymer on tissue. However, to impede adhesion or promote abhesion with tissue and blood components, surfaces with low energy must be created. A monolayer of closely packed trifluoromethyl groups on a surface has been shown to provide the lowest energy. (10) It is highly unlikely that such a monolayer could be easily created by synthesis or chemisorption on synthetic organic polymers. The chance of success is even more remote on the surface of living biological tissue where the condition is particularly unfavourable for condensation involving groups with replaceable hydrogen. Heparinized plastic surface delays the coagulation of plasma components, but adsorption of proteins still occur thereon (11). Studies on the biological tissue-polymer interface are being conducted by various research groups. The results of these studies are likely to provide useful information on the possibility of controlling adhesion and abhesion at this interface in various biomedical applications of the synthetic material.

REFERENCES

1. I. Redler, J. Bone and Joint Surg., 44, 1621 (1962)
2. W. F. Bernhard, A. S. Cummin, P. D. Harris, and E. W. Kent, Circulation, 27, 739 (1963)
3. G. M. Brauer and E. F. Huget, "The Chemistry of Biosurfaces," p. 781, vol. 2, M. Hair, ed., Marcel Dekker, N.Y. (1972)
4. E. Llewellyn-Thomas, P. Y. Wang, and J. S. Cannon, Polymer Preprints, ACS Publications, 33, 303 (1973).
5. E. Llewellyn-Thomas, P. Y. Wang, and N. Vinals J., Biomat. Med. Dev. Art. Org., 1, 507 (1973)
6. J. D. Galligan, F. W. Minor, and A. N. Schwartz, "Adhesion in Biological Systems," p. 258, R. S. Manaly, ed., Academic Press, N. Y. (1970)
7. J. Hollander, R. B. Gosnell, F. D. Trischler, and E. S. Harrison, Abstract of Papers, the 152nd ACS National Meeting, Paper K18 (1966).
8. J. W. Baker and J. B. Holdsworth, J. Chem Soc., 713 (1947).
9. C. W. Cooper, G. A. Grode, and R. D. Falb, "Tissue Adhesive in Surgery," T. Matsumoto, ed., p. 192, Med. Exam. Publishing Co., Flushing, N.Y. (1972)
10. R. E. Bair, E. G. Shafrin, and W. A. Zisman, Science, 162, 1360 (1968).
11. J. L. Brash and D. J. Lyman, "The Chemistry of Biosurfaces," p. 213, vol. 1, M. Hair, ed., Marcel Dekker, N. Y. (1971).

MEASUREMENTS OF THE QUANTITY OF MONOMER LEACHING OUT OF ACRYLIC BONE CEMENT INTO THE SURROUNDING TISSUES DURING THE PROCESS OF POLYMERIZATION

H. G. Willert, H. A. Frech
Orthop. Univ. Klinik
6 Frankfurt/M
Marienburgstrasse 2, West Germany

A. Bechtel
Battelle Institut e.V.
6 Frankfurt/M
Am Römerhof 35, West Germany

For the implantation of artificial joints into bones self-curing acrylic resins are widely used. This plastic material is available from different manufacturers as CMW-Bone Cement, Palacos, Simplex etc. A new development of the Swiss firm SULZER Broth. is Sulfix 6 and 9. Sulfix 6 and 9 were preliminary materials. Sulfix 6M is the definitive one untested by us as yet. Chemically these materials consist of methylmethacrylate or, in the case of Sulfix, of a mixture of methyl- and butyl-methacrylate. There are two components, a powder and a liquid. The powder consists of fully polymerized acrylate in form of small spherical granules and, in addition, of a milled "flour" form of acrylate. It also contains an activator, for instance benzoyl peroxide. The fluid contains the monomer of acrylate together with a stabilizer (hydrochinon) and an initiator (dimethyl paratoluidin) (Charnley 1970). Immediately before use the powder and the liquid are brought together in a ratio of 2:1 or in the case of Sulfix 6 and 9, 2.5:1. The monomer quickly dissolves the acrylic "flour" and the surface of the acrylic granules, while at the same time polymerization begins.

The resulting paste is very soft in the beginning

and changes to a rubbery mass, which can be molded and formed, within 1 1/2 to 4 minutes. After another 5-7 minutes it hardens completely. In the state where it can still be formed the bone cement is inserted and modelled into the bone. It is the drawback of this procedure that the bone cement sets completely only within the living organism. Therefore, the effects of polymerization -- the generation of heat and the evaporation of monomer -- can damage the surrounding tissues. In consequence of these adverse influences bone, bone marrow and connective tissues in the close surrounding of the cement become necrotic immediately after insertion in a zone up to 3 mm broad. The rise in temperature within the bone cement (reaching more than 70° C) have been the object of extensive studies published in the recent literature (Charnley 1970, Lautenschlager et al. 1973, Meyer et al. 1973, Ohnsorge et al. 1969 and 1970, Wiltse et al., 1957). On the other hand informations on the evaporation of monomer into the surrounding tissue are still very scarce and incomplete.

We, therefore, carried out several series of investigations to detect the amount of monomer leaching out of the cement. It is justified to assume that the main portion of monomer evaporates from the cement during the process of polymerization. After hardening residual monomer may remain within the polymerized bone cement (1.5-5%), but only minimal quantities still reach the surrounding tissues (Smith and Bains, 1956).

The following substances were tested: CMW-Bone Cement, Simplex, Palacos E and Palacos R, Sulfix 9.

The original Palacos, which had been used in the majority of cases investigated by Willert and Puls (1971) was no longer on the market. Instead of it Palacos E was tested in the same ratio of the mixture.

We thought it of great importance to carry out the investigations under conditions which could be reproduced easily. As a test tissue we, therefore, used human bone marrow as it was removed from the femur in the process of preparing the shaft for the implantation of the stem of the femoral prosthesis.

The components of bone cement were mixed in a ratio of 10g of powder to 5.2 ml of liquid at a constant temp. of about 22°C. They were stirred outside of the tissue medium to allow polymerization. Sulfix was mixed in a ratio of 10g of powder and 4g of fluid. The cement dough was then inserted into 15g of tissue medium contained in a test-tube. The test-tube was closed immediately by 2 paraffine films, whilst the cement inside hardened completely. In a water bath it was brought to a temperature of 37°C. Until it was analyzed up to 60 hours after polymerization the test material was stored in the refrigerator at +4°C.

We assumed that the monomer would have a greater affinity towards fat than towards albumen and water. We therefore separated the tissue medium bv the centrifuge (30 minutes at 5000 rotations/minute) before analysis into four fractions: 1. pure fat, 2. + 3. bone marrow fibres, -cells and other fatty substances and 4. red blood cells. In most instances fraction 2 was so small that it could not be separated with certainty from fraction 3. In these cases both fractions were analyzed together.

After separation the fractions were weighed, extracted with 3 ccm of dimethyl formamide (DMF) shaken and again given into the centrifuge for 20 minutes at 5000 rotations/min. To a defined volume of the DMF-phase (between 2 and 2.5 ccm), 5 mg (5 ml) of butyl acetate were added as internal standard. Then in each fraction the quantitative analysis of its content of monomer was carried out by gas chromatography.

Details of the Gaschromatographic Procedure

GC Model: Perkin Elmer dual-column gas chromatograph, model F7 with glass-coated injection port.
Recorder: Perkin Elmer Kompensograph 129-19 (2,5 mV)
Integrator: Perkin Elmer D 26
Separation column: 1.5 m glass column, i.d. 2.3 mm
Column packing: 10% Carbowax 20 M on kieselgur 70-100 mesh, "silanised" (E. Merck, Darmstadt)
Carrier gas: Nitrogen "special" (99.995%), 40 cm^3/min.
Detector: Dual-flameionisation detector (FID) in dual-circuit arrangement
Temperature of detektor: 180° C
Temperature of oven: 60° C

Injection port: 150°C
Burn gases: Hydrogen 50 cm^3/min, air 290 cm^3/min
Reagents: N,N-Dimethylformamide, puriss. (Fluka, Buchs, Switzerland), Butylacetate (puriss.)
Test substance: Methylmethacrylate, pure (Fluka AG)

For the gas-chromatographic determination, 0.5 µl of the sample was injected under the conditions described above.

The peaks were correlated by determination of their relative retention time r_s related to the internal standard.

Component	r_s
Methylmethacrylate	0,70
Butyl acetate	1,00
Dimethyl formamide	2,90

Calculation of the Monomeric Part of Methylmethacrylate of the Samples

The quantitative estimate was made by the internal standard method. The peak areas corresponding to methylmethacrylate and butyl acetate were determined by an electronic integrator. The concentration of monomer was determined by the ratio of the peak area of methylmethacrylate to the peak area per milligram butyl acetate (internal standard).

% Methylmethacrylate per weight of sample =

$$\frac{\text{mg BA} \cdot \text{area MAM} \cdot f_{sti} \cdot 100 \cdot 3}{\text{area BA} \cdot \text{weight of sample} \cdot X_{DMF}}$$

BA = butyl acetate
MAM = methyl methacrylate
f_{sti} = standard correction factor
X_{DMF} = volume of dimethylformamide taken after extraction.

The standard correction factor was determined as follows: 5 mg methylmethacrylate and 5 mg butyl acetate were weighed accurately and dissolved in 3 cm^3 dimethylformamide. 0.5 µl of this solution was gaschromatographed under the conditions described above. The standard correction factor f_{sti} was calculated according to the following equation:

$$f_{sti} = \frac{\text{mg methylmethacrylate . area butyl acetate}}{\text{mg butyl acetate . area methylmethacrylate}}$$

Evaporation of Monomer into the Tissue After Previous Polymerization for 90 seconds

In the first series of tests the cement dough was polymerized 90 seconds in the air before its insertion into the tissue medium. The results are shown in Table I.

After a given time of previous polymerization outside of the tissue medium very different amounts of monomer leach out from the individual types of cement into the tissue medium. We concluded that the reason for this behavior is their different speed of polymerization. Whereas CMW-Bone Cement at 22°C room temperature reaches the consistency suitable for insertion relatively soon after 90-120 seconds and then very quickly hardens completely, Sulfix 9 needs much more time to go through this process at the same temperature. The other cements were in between these two extremes. As the liquid free monomer is bound only in the process of polymerization, the longer it takes the cement to harden, the more monomer will evaporate. To be able to compare the different types of cement we, therefore, have to consider the characteristics of their polymerization:

Investigation of the Characteristics of Polymerization for the Different Types of Cement

To investigate the process of hardening of the different types of cement under standard conditions we used an apparatus called "penetrator". In principle the test gives measurements of the depth of penetration of a needle into the cement during hardening. The measuring standard for the depth of penetration was 0.10 mm. The needle could be loaded with different weights. The time of penetration was limited by a clock. The clock stopped the further penetration of the needle at the previously fixed time interval. After the penetration is stopped the depth of penetration can be read from a scale. The measurements were taken continuously during the process of polymerization at time intervals of 30 seconds. The cement used contained 5.2 ml of liquid and 10g of

TABLE I

Weight Percentages of Monomeric Methylmethacrylate in the Tissue Medium after 90 Seconds of Previous Polymerization

Cement	Fraction	Fat	Bone marrow fibres and cells	Red blood cells
CMW-Bone	Cement	1,3	0,7	0,02
		1,4	0,2	0,02
		0,9	0,9	0,01
		0,7	0,6	0,02
Simplex		2,64	0,47	0,16
		1,71	0,94	0,12
Palacos E		2,5	1,6	0,04
		2,5	2,0	0,02
		3,0	0,6	0,04
Sulfix 9	1.*	3,06	1,02	0,11
	2.*	1,59	0,23	0,003
	1.*	2,07	0,27	0,09
	2.*	0,78	0,05	0,0
Palacos R		5,1	0,2	0,04
		3,8	0,9	0,03
		4,1	0,7	0,08
		2,8	1,0	0,1

1.* Methyl-methacrylate
2.* Butyl-methacrylate

powder (4g of liquid and 10g of powder in the case of Sulfix 9). These measurements do not represent absolute degrees of hardness, but they are relative values of the depth of penetration per time unit and weight of the penetrating needle. They enable us to compare the processes of hardening of the different types of cement.

Of each type of cement mentioned above we tested the depth of penetration in at least 3 specimens and calculated the mean values for each of them. These values were transferred to a system of coordinates and produced curves which were characteristic for each type of cement. The curves altered their course by changes at the room temperature. The curves shown in Fig. 1 are representative for room temperatures of 22-23°C. The consistency most suitable for implantation is reached when the depth of penetration measures 10 mm. This grade of consistency was reached by CMW-Bone Cement in 116 seconds, by Palacos R after 210 seconds, by Palacos E after 219 seconds, by Simplex after 225 seconds and by Sulfix 9 after 470 seconds.

Quantity of Monomer Leached Out into the Tissue Medium After Previous Polymerization Until a Constant Degree of Consistency is Reached (10 mm Depth of Penetration)

In a further series of tests instead of a fixed time constant of 90 seconds for the previous polymerization outside of the tissue medium, a consistency of 10 mm depth of needle penetration was chosen as the constant factor. That means that CMW-Bone Cement was inserted 116 seconds, Palacos R 210 seconds, Palacos E 219 seconds, Simplex 225 seconds and Sulfix 9 470 seconds after mixing the liquid and powder components. All other conditions of the experiment remained the same. The results are shown in Table II.

The quantities of monomer which evaporated into the surrounding tissue during the process of polymerization found in the different fractions are now fairly the same for all types of cement. However all values are also considerably smaller because the cements were brought into contact with the tissue medium 25-380 seconds later than in the first series. Our assumption, that the great differences in TABLE I are caused by the different speed of polymerization

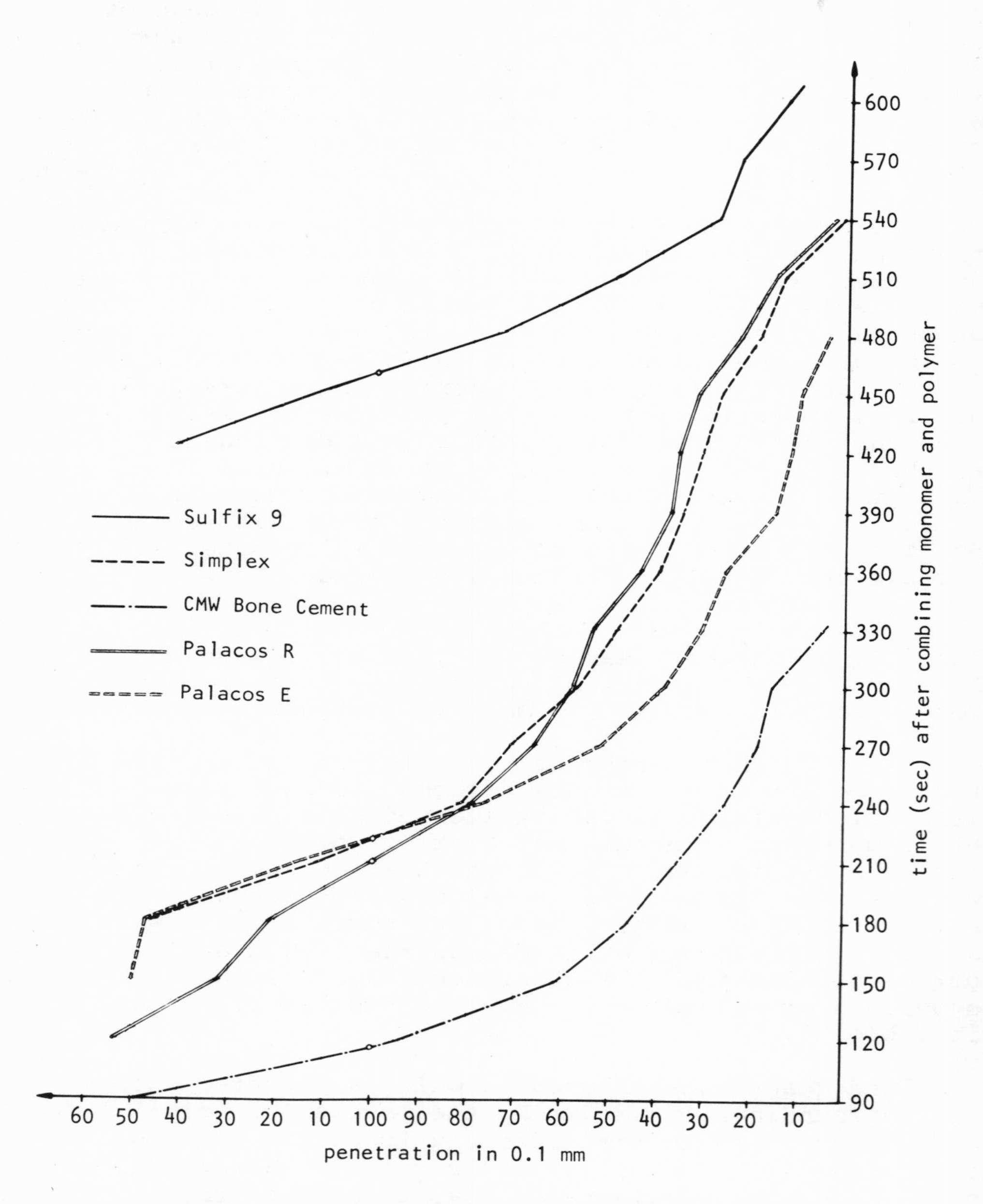

Fig. 1. Hardening of Autopolymerizates. Room temperature -22.5°

TABLE II

Weight Percentages of Monomeric Methylmethacrylate After Previous Polymerization up to a Consistency which Allowed 10 mm Depth of Needle Penetration

Cement	Fraction	Fat	Bone marrow fibres and cells	Red blood cells
CMW-Bone Cement		1,05	0,02	0,02
		0,48	0,38	0,04
Simplex		1,50	0.6	0,5
		0,86	0,27	0,17
Palacos E		0,7	0,09	0,03
		0,6	0,1	0,06
Sulfix 9	1.*	0,45	0,02	0,02
	2.*	0,22	0,0	0,0
	1.*	0,23	0,03	0,02
	2.*	0,10	0,0	0,0
Palacos R		0,8	0,3	0,03
		0,7	0,06	0,04

1.* Methyl-methacrylate
2.* Butyl-methacrylate

of the individual types of cement has been confirmed by these findings. Once polymerization has progressed to a consistency of 10 mm of penetration, further evaporation of monomer is fairly of the same quantity in all types of cement used in these experiments.

DISCUSSION

In our tests allowing a previous polymerization outside of the tissue medium before insertion of the cement into the tissue for 90 seconds or at a depth of penetration of 10 mm the amount of monomer found in the tissue shows marked differences for the different fractions. The highest content of monomer was always found in the fat, the lowest in the red blood cells. This confirms our assumption that monomer of methylmethacrylate has a definite affinity towards the fatty substances of human tissue. The monomer of methylmethacrylate is highly cytotoxic (Hulliger, L. 1962). According to the investigations of Mohr (1958) the monomer kills the tissue cells in a concentration of 1:1000 and upward. This necrotizing effect becomes more and more marked with rising concentrations of monomer. Together with May we investigated the cellular toxicity of monomeric methylmethacrylate in cell cultures of amnion cells and human embryonic fibroblasts. We found already moderate signs of inhibition of cell growth in a dissolution of 1:1000, which is equal to a concentration of 0,001 weight %. In a solution of 1:100 (0.01 weight %) the inhibition of growth was very marked. It appears from these findings that the concentration of monomer as measured by gas chromatography reaches cytotoxic levels in all fractions of the tissue medium. This seems to prove the toxic effect of methylmethacrylate monomer on the surrounding tissue of the implant bed.

However the importance of the initial damage to the tissue should not be overrated, for our morphological investigations of the bony implant bed (Willert and Puls 1971) show that after a few months, at least after 2 years, the necrotic areas are being replaced by living tissue. So one can assume with some certainty that the damage to the tissues caused by the implantation will be repaired by the organism in the course of time.

In the long run, however, changes in the cement caused by its contact to the living tissues may have adverse effects. We conclude this from the fact that monomer is lost into the bed of the implant during polymerization. It might be possible that this loss interferes with the ratio of polymer and monomer in the superficial layers of the cement in a way that polymerization remains incomplete and the connection between the globules remains insufficient. The cells of the surrounding tissue (foreign body giant cells as a rule) now have a chance to invade the cement in between the globules and it might be possible for them to crack their formations. Although we still have to consider this as a hypothesis it can explain our morphological findings, which show uni- and multinucleated cells, in places accompanied by blood vessels, which have invaded the spaces between the globules of the bone cement.

CONCLUSIONS

For surgical practice from our investigations we derive the recommendation to insert the bone cement into the tissues not too early. As more and more monomer is bound into the cement mass during polymerization, it is advised to let the material harden outside of the body to a consistency which just still allows it to model closley to the bone surfaces. According to Charnley (1970) and Lee and co-workers during this phase monomer already evaporates from the superficial layer of the acrylic mixture. Sufficient polymerization outside of the body considerably reduces the toxic effect to the tissues and the hazards of loss of monomer from the acrylic mixture. For the same reason it is not advisable to attempt to absorb the heat caused by polymerization by using a prosthesis cooled beforehand in the refrigerator. This only slows down the process and opens a chance for increased evaporation of monomer within the organism.

ACKNOWLEDGEMENT

The authors wish to thank Roehm GmbH, Darmstadt and Wienand Sohne & Co., Gmbh, Sprendlingen for their cooperation during the development of the methods.

BIBLIOGRAPHY

Charnley, J. Acrylic Cement in Orthopaedic Surgery, E & S Livingstone, Edinburgh and London 1970.

Hulliger, L. Untersuchungen über die Wirkung von Kunstharzen (Palacos und Ostamer) in Gewebekulturen Arch. orthop. Unfall-Chir. 54, 581 - 588 (1962).

Lautenschlager, E. P., Moore, B. K., and Schoenfeld, C.M. Physical Characteristics of Setting of Acrylic Bone Cements 5. Annual Biomaterials Symposium Clemson/S.C. 1973.

Lee, A. J. C., Ling, R. S. N. and Wrighton, J. D. Some properties of polymethylmethacrylate with reference to its use in orthopaedic surgery, Clin. Orthop. 95, 281 (1973).

May, G. Personal communications.

Meyer, P. R., Lautenschlager, E. P. and Moore, B. K. On the Setting Properties of Acrylic Bone Cement J. Bone & Jt. Surgery 55A, 149 - 156 (1973).

Mohr, H. J. Pathologische Anatomie u. kausale Genese der durch selbstpolymerisierendes Methakrylat hervorgerufenen Gewebsveränderunger Z. ges. exp. Med. 130, 41 - 69 (1958).

Ohnsorge, J., Goebel, G. Oberflächentemperaturen des abhärtenden Knochenzementes Palacos beim Verankern von Metallendoprothesen im Oberschenkelmarkraum. Arch. orthop. Unfall-Chir. 67, 89 - 100 (1969).

Ohnsorge, J., Goebel, G. Die Verwendung unterkühlter Metallendoprothesen in der Hüftchirurgie. Z. Orthop. 107, 683 - 696 (1970).

Ohnsorge, J., Kroesen, A. Thermoelektrische Temperaturmessungen des abhärtenden Knochenzementes "Palacos". Z. Orthop. 106, 476 - 482 (1969)

Smith, D. C. and Bains, M. E. D. The detection and Estimation of Residual Monomer in Polymethylmethacrylate. J. D. Res. 35, 16 (1956).

Willert, H. G. und Puls, P. Die Reaktion des Knochens auf Knochenzement bei der Allo-Arthroplastik der Hüfte. Arch. orthop. Unfallchir. 72, 33 - 71 (1971).

Wiltse, L. L., Hall, R., Stenehjem, J. C. Experimental Studies Regarding the Possible Use of self-curing Acrylic in Orthopaedic Surgery. J. Bone Jt. Surg. 39A, 961 - 972 (1957).

MODIFICATION AND CHARACTERIZATION OF POLYSTYRENE SURFACE USED FOR CELL CULTURE

T. Matsuda and M. Litt

Division of Macromolecular Science
Case Western Reserve University
Cleveland, Ohio 44106

INTRODUCTION

An interest has developed in the biomedical applications of polymers. This includes problems related to the interaction of biosurface and polymer surface. Cell culture is remarkably affected by the surface of the culture vessel (1-3). The surface factors considered to be important in cell culture are:

1. concentration of polar-nonpolar group (hydrophilic-hydrophobic balance)
2. concentration of ionic groups
3. nature of the polor and nonpolar groups.

Polystyrene petri dishes have been commonly used for cell culture. Polystyrene as a hydrocarbon has a low surface energy and is classified as a non-specific adsorbent (4). This leads to its poor wettability by water and to very low adsorptive and adhesive properties with amniotic fluid cells. In order to improve the properties of polystyrene surfaces, polar sites were created at the surface of the polymer by sulfonation. The surface properties of chemically modified polystyrene were investigated from the point of view mentioned above.

EXPERIMENTAL

Materials. The samples for this study were

polystyrene petri dishes (Falcon 1007) with a bottom surface area of 21 cm^2. The 5% fuming sulfuric acid (based on SO_3) was prepared from sulfuric acid (sp.gr. 1.84, pure grade) and commercially available 30% fuming surfuric acid (reagent grade). The radioactive $^{45}CaCl_2$ was obtained from Amersham/Searle Corp.

Surface Modification. The petri dish was dipped into 5% fuming sulfuric acid at room temperature for a given time, washed several times with distilled water and then air-dried. The sulfonated polystyrene surfaces carrying sodium, magnesium, calcium and aluminum cations were prepared by dipping the dish into an aqueous solution of sodium hydroxide (10% wt.), magnesium chloride (10% wt.), calcium chloride (10% wt.) or aluminum acetate (saturated solution), respectively. In a similar manner, the organic amine and ammonium salts of sulfonated surfaces were prepared from the aqueous solution (10% wt.) of the corresponding amine; in addition, two polymers, polyethylenimine and protamine sulfate, were coated on the sulfonated surface by the same technique (concentration; about 10% wt.). All such surfaces were washed three times with distilled water.

Contact Angle Measurements. The contact angles toward water of the surfaces prepared above were measured at constant humidity and temperature (50% r.h. & 70°F) using tridistilled water by means of either the light beam reflection technique (5,6) or the sessile drop technique (NRL Contact Angle Apparatus manufactured by Hart-Ramey Corp.). The standard deviation of each measurement was about 2°.

Surface Conductivity Measurements. Surface conductivities of treated surfaces were determined at constant humidity and temperature (50% r.h. & 70°F) with a Keithley 610B Electrometer by using copper electrode and silver-filled epoxy resin ad adhesive agent, according to ASTM Standards for Plastics, D257.

Adsorption Measurements. To determine the surface density of the polar sites created on the surface of polystyrene petri dish by sulfonation, the radioactive technique was used. An aqueous solution of ^{45}Ca labelled calcium chloride (2.9×10^{-6} g/ℓ) was placed on the surface of the sample; the radioactive solution was removed, and the surface was rinsed with absolute methanol several times in order to remove the contaminating radioactivity of the aqueous solution adhering to the surface. After air-drying the radioactivities originating from the adsorbed ^{45}Ca ions on the surface were measured by using a Nuclear-Chicago Gas Flow Detector (Model 480) operated in the Geiger Counting mode and a Radiation Instrument Development Laboratory Scaler (Model 49-23). The specific activity of calcium ions necessary for the calculation of the amount of calcium ions adsorbed was measured at the same time under the same geometrical conditions as for the standard sample, which was prepared by evaporating a known amount of the radioactive sample solution deposited on the surface. The specific activity of the radioactive $CaCl_2$ under the experimental geometry was 1.30×10^{-14} g/cm^2 c.p.m. The reproductivity of the concentration of polar sites by this technique is shown in Table I. Under these experimental conditions the radioactivity of an unsulfonated polystyrene treated as above was about 95 c.p.m., which was 3% of the radioactivity of the least sulfonated surfaces

RESULTS AND DISCUSSION

Under the usual conditions for preparation of cationic-exchange resin using 30% fuming sulfuric acid or chlorosulfuric acid, the polystyrene surface was immediately degraded, while in the present experiment using 5% fuming sulfuric acid the surface appeared unchanged. In order to understand the structure of the treated surface, two different spectroscopic methods were used. X-ray photoelectron spectroscopy which can determine the chemical composition of the surface within 50Å depth (7), showed the existence of sulfur and oxygen atoms in addition to carbon atoms, but the concentrations of the former two atoms detected were very small compared with carbon atoms. Attenuated total reflection infrared spectra (ATR) of the modified surfaces were also obtained. As is well known, ATR techniques permit the recording of spectra of films a few molecular layers thick (8). The surface treated by sulfuric acid showed weak to medium bands at 1100 and 1200 cm^{-1} ascribable to the sulfonic acid group (9), they appear only after sulfonation. From these results, it was concluded that moreover reactions to introduce sulfonic acid group took place only at the outermost bonds in the polymer surface or in a thin layer which may be no thicker than a single molecule.

TABLE I

Reproducibility of the Concentration of Polar Sites on Sulfonated Polystyrene

Sample No.	Radioactivity (c.p.m.)	^{45}Ca-ion (g/cm^2 x 10^{10})	^{45}Ca molecule /A^{o2} (x10^4)	10^3_A ^{o}Sites
(A) Polystyrene(Falcon 1007) treated with 5% fuming sulfuric acid for 5 min.				
1	3387	0.44	0.59	17
2	3572	0.47	0.62	16
	Av.3480	0.46	0.61	16
(B) Polystyrene (Falcon 1007) treated with 5% fuming sulfuric acid for 1 hr.				
3	31125	4.05	5.4	1.85
4	30874	4.02	5.4	1.87
5	32610	4.24	5.7	1.77
6	31563	4.11	5.5	1.83
7	31641	4.18	5.5	1.82
	Av.31563	4.12	5.5	1.83
(C) Polystyrene (Falcon 3002) treated with 5% fuming sulfuric acid for 1 hr.				
8	Av.31563	4.12	5.5	1.83

Wettability. The treated polystyrene surfaces appear unchanged to the eye but now are wetted more easily than the untreated one. The advancing and receding contact angles, which depend on the concentration of polar and ionic groups, are a measure of wettability of the substrate surface. As can be seen in Fig. 1, the water contact angle drops as a function of sulfonation time to where the advancing angle reaches to 31°, while the receding angle reaches to 0°. The water contact angle of the treated surfaces, Table II, carrying inorganic cations (Na^+, Mg^{++} and Al^{+++}) showed these are more wettable than the sulfonated surface itself (H^+ cation). Thus, the exchange of protons for Na^+, Mg^{++}, Ca^{++} or Al^{+++} improves the hydrophilic character of these surfaces.

Surface Conductivity. The surface conductivity of the chemically modified polystyrene is another measure of the polarity arising from the chemical modification and depends to a large extent upon the presence of ionic groups on the surface. Table II shows the decrease in resistivity that occurred with sulfuric acid-treated polystyrene surface. The sulfonated surface was made more polar and conductive when metallic cations replaced the proton. This is very unusual and may be due to adsorption of water by the polyvalent cations. Thus, the results of both contact angle and surface conductivity measurements show that the polarity of the surface can be appreciably increased, and that a high-energy relatively polar surface is created.

Adsorption and Desorption of Radioactive ^{45}Ca Ions. The extent of chemical modification, e.g. the surface density of the ionic sites, was determined by the adsorption of calcium ion using radioactive techniques. The adsorbed calcium ions and the surface density of polar sites are plotted in Fig. 2 as a function of the sulfonation time. A relatively low concentration of polar sites was formed, one site per 16,000A°^{2} after 5 minutes of sulfonation rising to one site per 1,800A°^{2} after 60 minutes. The extent of chemical modification was almost proportional to the sulfonation time under the present experimental conditions. These results help confirm the hypothesis that the sulfonation occurred in the superficial surface layer only. The wettability by water of the sulfonated surfaces was proportional to the amount of polar sites produced, and therefore the changes in contact angle are relevant only to the modification of the surface (Fig. 3).

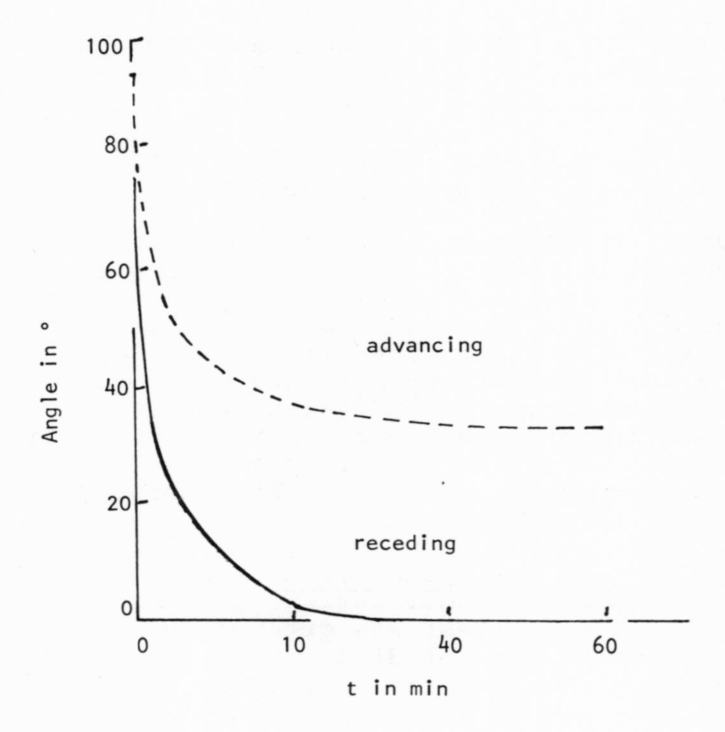

Fig. 1. Variation of water contact angle with changes in sulfonation time.

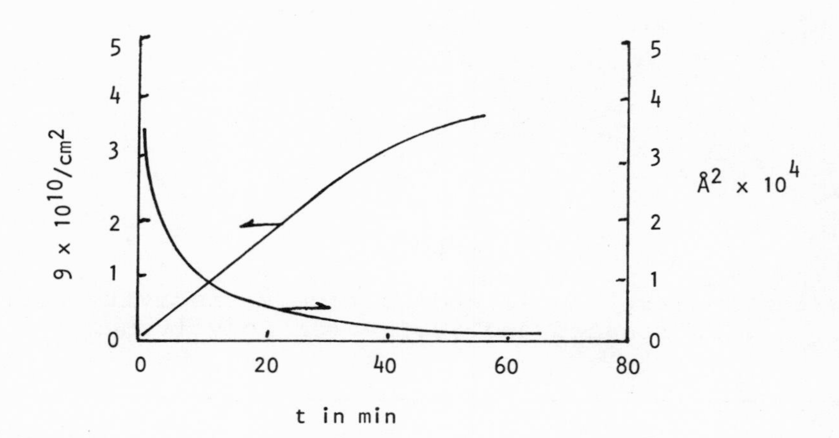

Fig. 2. Change in the degree of chemical modification with sulfonation time in min. for grams of $^{45}CaCl_2$ in $9 \times 10^{10}/cm^2$ and for average area per polar site in $Å^2 \times 10^4$.

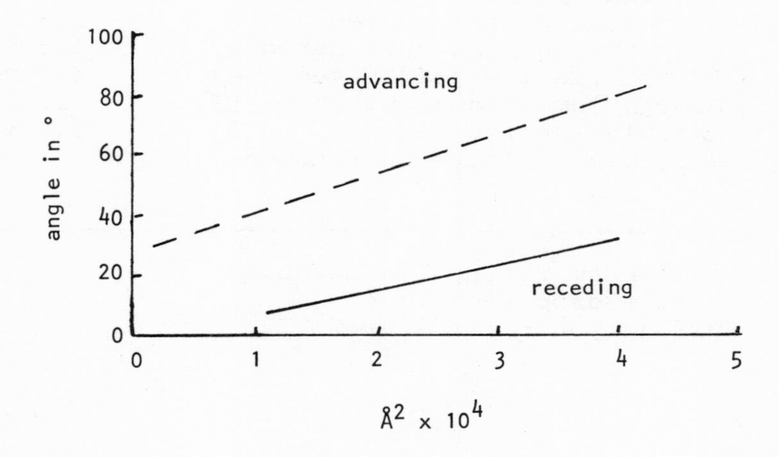

Fig. 3. Water contact angle vs. surface density of polar sites, expressed in $Å \times 10^4$ per site.

TABLE II

Water Contact Angle and Surface Resistivity of Sulfonated and Postreated Polystyrne

Code Name	Treatment***	Water Contact Angle* (°)	Surface Resistivity ohm-square
K-12	nontreated	>90(94**)	1.9×10^{14}
K-1	sulfonated	31	7.3×10^{11}
K-2	Na-sulfonated	22	9.0×10^{9}
K-3	Mg-sulfonated	20	2.5×10^{9}
K-4	Ca-sulfonated	18	4.5×10^{9}
K-5	Al-sulfonated	15	1.4×10^{9}
K-6	Ethylenediamine-sulfonated	28	6.5×10^{10}
K-7	Ethanolamine-sulfonated	30	1.8×10^{11}
K-8	Polyethylenimine-sulfonated	23	1.5×10^{10}
K-9	Trimethylamine-sulfonated	31	4.7×10^{12}
K-10	Dimethylamine-sulfonated	29	1.2×10^{11}
K-11	Ammonium-sulfonated	25	2.5×10^{10}
K-13	Protamine sulfate-Sulfonated	27	-

*) Light beam reflection technique
**) Sessile drop technique
***) Sulfonation time; 1 hr.

The ease of desorption of ^{45}Ca ions from the surface by various aqueous solutions is given in TABLE III. [All exchange experiments were conducted on surfaces which had been sulfonated for one hours.] Rinsing with distilled water gradually decreases the adsorbed calcium ions, while almost all of the calcium ions are removed instantly by rinsing with 0.2 mol. aqueous ionic solutions such as nonradioactive calcium chloride or disodium ethylene-diaminetetraacetate. The desorption behavior shows that the major part of the adsorption of calcium ions is ion-exchange adsorption at the surface, since they can be removed very rapidly.

Cation-exchange Capacity of Sulfonated Surface. The treated surfaces [one hour sulfonation] which had been conditioned with

1. Na^+ solution
2. Mg^{++} solution
3. Ca^{++} solution
4. Al^{+++}solution

were contacted with 6.4 x 10^{-8}M radioactive $CaCl_2$ solution, and surface radioactivity was measured as a function of time. Fig. 4 shows the variation of degree of ion-exchange with immersion time in the radioactive $CaCl_2$ solution. The degree of ion-exchange was calculated by dividing the measured radioactivity by the radioactivity of the sulfonated surface containing only radioactive calcium ions. As can be seen in Fig. 4, the order of exchange rate was Na+>>Mg++> Ca++>>Al+++, and this order agrees with the ion-selectivity sequence of sulfonated polystyrene ion-exchange resin (10). The Al+++ salt surface did not exchange under these conditions. Such studies can tell us what happens to treated surfaces under physiological conditions with various ions present. The first-order plots for Ca++ and Mg++ with respect to exchange time are shown in Fig. 5. Na+-sulfonated surface showed a fairly linear first-order plot almost up to 90% completion of ion-exchange with a rate of 0.17/min., while for Ca++ and Mg++ -sulfonated surfaces, both first-order plots seem to consist of two straight lines. These may imply a heterogeneous distribution of polar sites; that is, there are two different types of polar sties on the surface, one consists of isolated polar sites, while the other consists of associated sites. The sulfonated surface coated with polyethylenimine, which is a polymeric salt had no cation-exchange capacity under the low concentration of calcium ion. Even exposure to 0.1 M $CaCl_2$ for 20 min. did not remove the polymer at the surface.

TABLE III

Desorption and Removal of $^{45}Ca^{++}$ From the Surface of Sulfonated Polystyrene by Various Sources

Sample No.	Initial Radioactivity (c.p.m.)	Treatment	Remaining Radioactivity (c.p.m.)***
4	30874	0.2 M EDTANa* aqueous solution, 30 min	253
5	32610	0.2 M $CaCl_2$ aqueous solution, 1 min	377
3	31125	aqueous solution 30 min	276
6	31563	aqueous solution 60 min	320
7	31641	Distilled water	
		0.67 hr	28,261
		5.75 hr	6,567
		28 hr	3.704

*) EDTA-Na (Disodium ethylenediaminetetraacetate)
**) Background radioactive counts were 75 c.p.m.

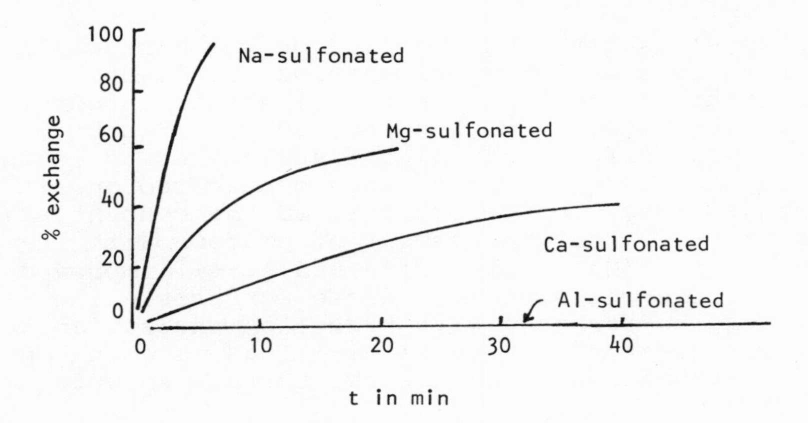

Fig. 4. Degree of Ion-exchange of ^{45}Ca (in %) with absorbed ions as a function of immersion time.

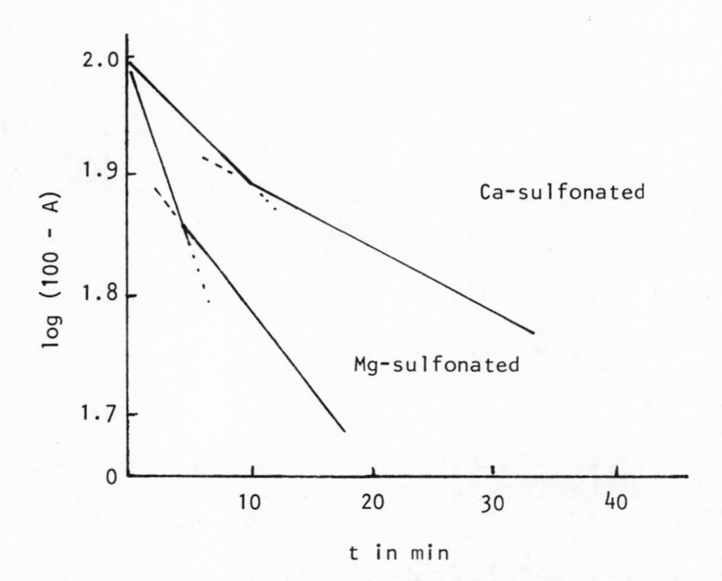

Fig. 5. First-order plot for exchange of ^{45}Ca-ions with absorbed ions where A is degree of ion-exchange, as a function of immersion time.

CONCLUSION

Polar sites have been created on the surface of polystyrene by a suitable sulfonation technique. Under given conditions, sulfonation of polystyrene produced a reproducible amount of high-energy surface ionic sites. These surfaces had higher surface energies and higher surface conductivities than nontreated ones. Their wettability was proportional to the concentration of polar sites. The replacement of protons with inorganic cations (Na^+, Mg++, Ca++ and Al+++) improved the hydrophilic characters of these surfaces. The order of cation-exchange with ^{45}Ca++ on the sulfonated surface was Na+>> Mg++> Ca++>> Al+++, as is found in normal ion-exchange resin, but the surface showed much greater selectivity.

ACKNOWLEDGEMENTS

The authors wish to acknowledge the support of this study by the National Institute of Health (Grant Number HD-00592-02).

REFERENCES

1. C. Rapport, J. P. Poole and H. P. Rapport, Exptl. Cell Res., 20, 465 (1960).
2. R. E. Baier and W. A. Zisman, Naval Res. Lab. Rept. 6755, 22 pp (1968).
3. L. Weiss and L. E. Blumenson, J. Cellular Physiol., 70, 23 (1967).
4. A. V. Kiselev, J. Colloid and Interface Sci., 28, 430 (1968).
5. I. Langmuir and V. J. Schaeffer, J. Am. Chem. Soc., 59, 2405 (1937).
6. T. Fort, Jr. and H. F. Patterson, J. Colloid Sci., 18, 217 (1963).
7. D. A. Shirley, Electron Spectroscopy, American Elsevier Publishing Company, Inc., New York, 1972.
8. P. A. Wilks, Jr. and T. Hirschfeld, Appl. Spectry. Rev., 1, 99 (1967).
9. L. J. Bellamy, The Infrared Spectra of Complex Molecules, 2nd, Ed., J. Wiley and Sons, Inc., New York, 1958, p. 350.
10. O. Bonner and L. Smith, J. Phys. Chem., 61, 326 (1957).

A NEW METHOD FOR THE PREPARATION OF NONTHROMBOGENIC SURFACE BY RADIATION GRAFTING OF HEPARIN: PREPARATION AND IN-VITRO STUDIES

A. S. Chawla and T. M. S. Chang

Department of Physiology
McGill University
Montreal, Quebec, Canada

SUMMARY

As the cellulose type of membranes are being used for hemodialysis, radiation grafting of heparin to these types of membranes was tried. Using radioactive heparin-^{35}S it was found that no detectable heparin could be bound to collodion or to cellulose acetate membranes. Washing with saline or with human plasma of heparin bound to the DEAECA membranes showed that 2 hour irradiated membranes retained the highest amount of heparin. However, there was a decrease in tensile strength by about 25% in these 2 hour irradiated membranes. Blood did not clot even after 60 minutes on the membranes in which heparin was bound by 1 or 2 hours of irradiation. *In-vitro* studies appear to be promising.

There are many ways to prepare nonthrombogenic surfaces (1), one of these is the attachment of heparin to surfaces. Since the preparation of heparinized surfaces by Gott and his co-workers (2,3), extensive research has been carried out to extend and develop this type of nonthrombogenic surface. Most of the methods developed, including the initial procedure of Gott et al (2,3), involve binding quaternary groups to the surface to be heparinized. Heparin due to its sulphate groups is negatively charged and hence complexes with the quaternary group on the material. Quaternary groups may be added to the surface by a variety of methods. The following are some examples: Yen and Rembaum (4) prepared polyurethanes containing pendant amino groups which was then quaternized. Falb et al (5,6) grafted styrene followed by treatment with dimethyl aniline and then quaternization of amine groups. Merker et al (7) used δ -aminopropyltriethoxysilane to react with silica filler in silicone rubber and the amino group is then quaternized. Simpler procedures have been developed by Grode et al (8,9), who treated polymers with a solution of tridodecylmethylammonium chloride (TDMAC). The hydrocarbon chain of the salt dissolves in the polymer and on removing the solvent,quaternary groups are left on the surface of the polymer. In a modification of this process (9), heparin-TDMAC complex is formed and then this complex is applied to the polymer. Lagergren and Erikson (10) have described a process where polymer is softened by heating in an aqueous solution of quaternary salt. Hydrocarbons part of the salt penetrate into the polymer and is fixed on cooling. To prevent leaching of heparin, it could be cross-linked by glutaraldehyde (11). Heparin may also be bonded covalently to some of the polymers. Merrill et al (12) prepared a cross-linked material by cross-linking heparin and polyvinyl alcohol by aldehydes. Halpern and Shibakawa (13) prepared polystyrene containing isocyanate groups to which heparin was covalently bound vis its hydroxl group.

Cellulose type of polymers have also been heparanized. Merrill et al (14) described a method for binding heparin to cellulose membranes through the use of polyethylene-imine. Martin et al (15) sorbed heparin on the membranes prepared from a blend of cellulose acetate and N, N-diethyl-aminoethyl cellulose acetate. In the present work, a new method (16) of binding heparin by gamma-radiation is described.

MATERIAL AND METHODS

Three types of membranes were used in the present study. Cellulose nitrate obtained from collodian (Fisher Scientific), was dissolved in acetone (15 gm in 100 ml acetone). Cellulose acetate (Eastman Kodak 40-25) was dissolved in acetone (18 gm in 100 ml acetone). Cellulose N, N-diethylaminoethyl ether (Eastman Kodak) was acetylated by reaction with acetic anhydride using $ZnCl_2$ as a catalyst (17). The reaction was stopped by adding water. The N, N-diethylaminoethyl cellulose acetate (DEAECA) formed was precipitated in water, washed and then dried in vacuum oven. DEAECA thus obtained, was dissolved in acetic acid (15 gms in 100 ml of 65% acetic acid).

The membranes were cast with a knife to glass plate clearance of 10 mil and then gelled in cold water. After washing thoroughly these were soaked in a water solution of 0.25g % heparin (sodium salt) tagged with heparin - ^{35}S (Calbiochem). The soaked membranes were then irradiated with gamma ray in the "Gamma Cell 220" (Canadian Atomic Energy Ltd.) operating at a dose rate of 2.4 x 10^5 rads per hour.

Three in-vitro tests were used to test how strongly heparin has been bound to the membranes: (i) In this test a 2" x 2" membrane piece (0.015 cm in thickness) was placed in 100 ml of normal saline. At given intervals the washing solution was changed and at the same time a membrane sample was taken to measure its heparin content. (ii) The membrane was clamped in a modified CRC dialysis cell with stainless steel wire mesh supports on both sides of the membrane. The membrane was washed on both sides with 100 ml of recirculating washing solution (normal saline or human plasma) at a rate of 100 ml/min. The washing solution was monitored for the heparin being washed out. Membrane samples were taken before and after washing for heparin content. (iii) Here 20 circular (diameter 0.78 cm) heparinized membranes were washed by soaking in 100 ml of washing solution (saline or plasma). For sampling, individual circular pieces of membrane were withdrawn from the solution at frequent intervals.

To find the activity (and hence the heparin concentration) in the membrane samples or in the plasma samples, these were dissolved in $HClO_4$ and H_2O_2 according to the procedure of Mahin and Lofberg (18). Counting was done in a Tri-Carb liquid scintillation counter (Packard Instrument Co. Model 3375).

Stress-strain relations were measured using an Instron testing machine. The membranes were kept under water for these measurements.

In-vitro Clotting Time Measurements

The following clotting time measurements were

carried out at 37°C in glass test tubes; in glass test tubes coated with cellulose membrane; or in glass test tubes coated with cellulose membrane and then heparinized. To each was added 1 ml of human ACD blood. To start the clotting time measurement, 0.1 ml of 0.1 M $CaCl_2$ solution was added.

In-vivo Testing

Heparinized surfaces were tested in dogs weighing about 25 kg. A test piece (App 0.5 mm x 12 cm long) could be inserted into the jugular vein to a length of about 10 cm through a 16 gauge needle. After insertion the needle was removed and the polymers were held in place by a hemostat. A control piece of polymer and a heparin grafted piece of polymer placed separately in one of the jugular veins of the same animal. At the end of two hours, 2.5 ml of 1g% heparin solution was injected intravenously into the dog to prevent further clot formation. The vein was then cut open along the length and the test piece was carefully removed and examined for clot formations on it.

RESULTS AND DISCUSSIONS

With radiation doses up to 3 x 10^6 rads no detectable heparin was bound to the cellulose nitrate membranes. Furthermore, with these radiation doses, the cellulose nitrate membranes tend to become brittle. This phenomenon of membranes becoming brittle results from the degradation of cellulosic chains (19). Similar results, in that no heparin could be radiation grafted, were obtained with cellulose acetate membranes. Successful results were obtained with DEAECA membranes and the details are as follows:

The results of non-irradiated, 1 hour irradiated and 2 hour irradiated membranes tested by method i are given in Figure 1. It is evident that in the beginning there is a significant decrease in the heparin content of the membrane as some heparin is washed out, but after 2 hours of washing further removal of heparin is negligible. Furthermore, a significant amount of heparin is left in the membrane in the 1 hour and 2 hour irradiated samples but the amount of heparin left on non-irradiated samples is very small. Non-irradiated membranes are those membranes which were heparinized by soaking in heparin

TABLE I

CONCENTRATION OF BOUND HEPARIN AFTER 6 HOURS OF WASHING BY METHOD II

Time of Irradiation of the membrane	Washing solution	Conc. of Heparin after washing ($\mu g/cm^2$)
0 hour	saline	15.7
	plasma	12.4
1 hour	saline	46.5
	plasma	30.0
2 hours	saline	61.0
	plasma	55.0

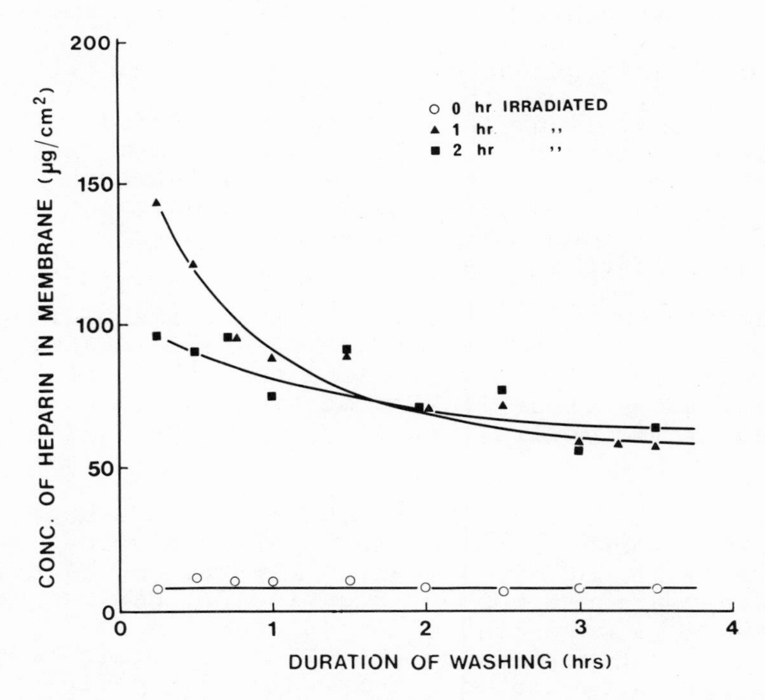

Figure 1 Washing of heparinized DEAECA membranes with saline.

solution but did not involve any irradiation.

Figures 2a and 2b show the results obtained by tests using method ii in which the amount of heparin coming out in the wash was measured. The concentration of heparin shown on the ordinate refers to the amount being washed out per cm^2 of the membrane area. Here again it is evident that in the case of irradiated membranes, after the initial rapid wash out of heparin, the level stabilizes. In the case of non-irradiated membranes, little heparin comes out in solution since the amount originally present in the membrane is negligible after the pretest washing. Table I gives the concentration of heparin left on the membranes after washing with saline and with plasma.

Figure 3 shows the results obtained by testing using method iii. Each point in the diagram represents one circular piece of membrane taken out as a sample. Again it is evident that heparin washes out rapidly at first and then it stabilizes. In each of these figures it is evident that 2 hour irradiated samples retained heparin the most. In the case of non-irradiated membranes, the heparin remaining in them is very small.

As cellulose chains are known to degrade by radiation (19), it was important to know how much degradation has taken place during these irradiations. This was found from tensile strength measurements shown in Figure 4. There is a linear relationship between the tensile strength and the duration of irradiation. 2 hours of irradiation resulted in about 25% decrease in tensile strength.

DEAECA binds heparin by its primary amino groups. It is postulated that these bonds result in DEAECA and heparin molecules lying very close to each other and this facilitates cross-linking of heparin not only to other heparin molecules, but also to the cellulose chains.

The results of in-vitro clotting studies show that DEAECA membranes soaked in heparin solution but not followed by any radiation may prolong clotting time if they are not washed extensively. However, if these membranes were washed for 48 hours in water, the clotting times reported in Table II were obtained. Blood in carefully washed non-irradiated heparin-soaked membranes clotted in 12.5 minutes, but blood in 1 and

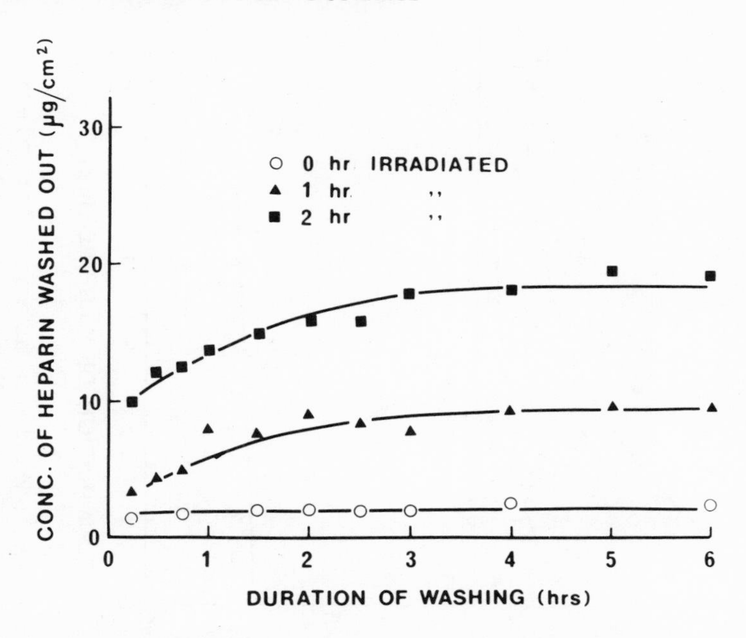

Figure 2A Washing of heparinized DEAECA membranes with saline.

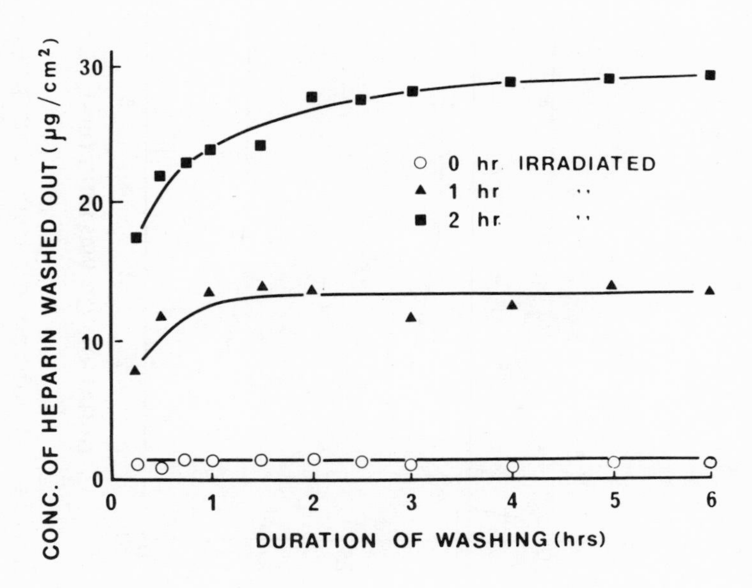

Figure 2B Washing of heparinized DEAECA membranes with plasma.

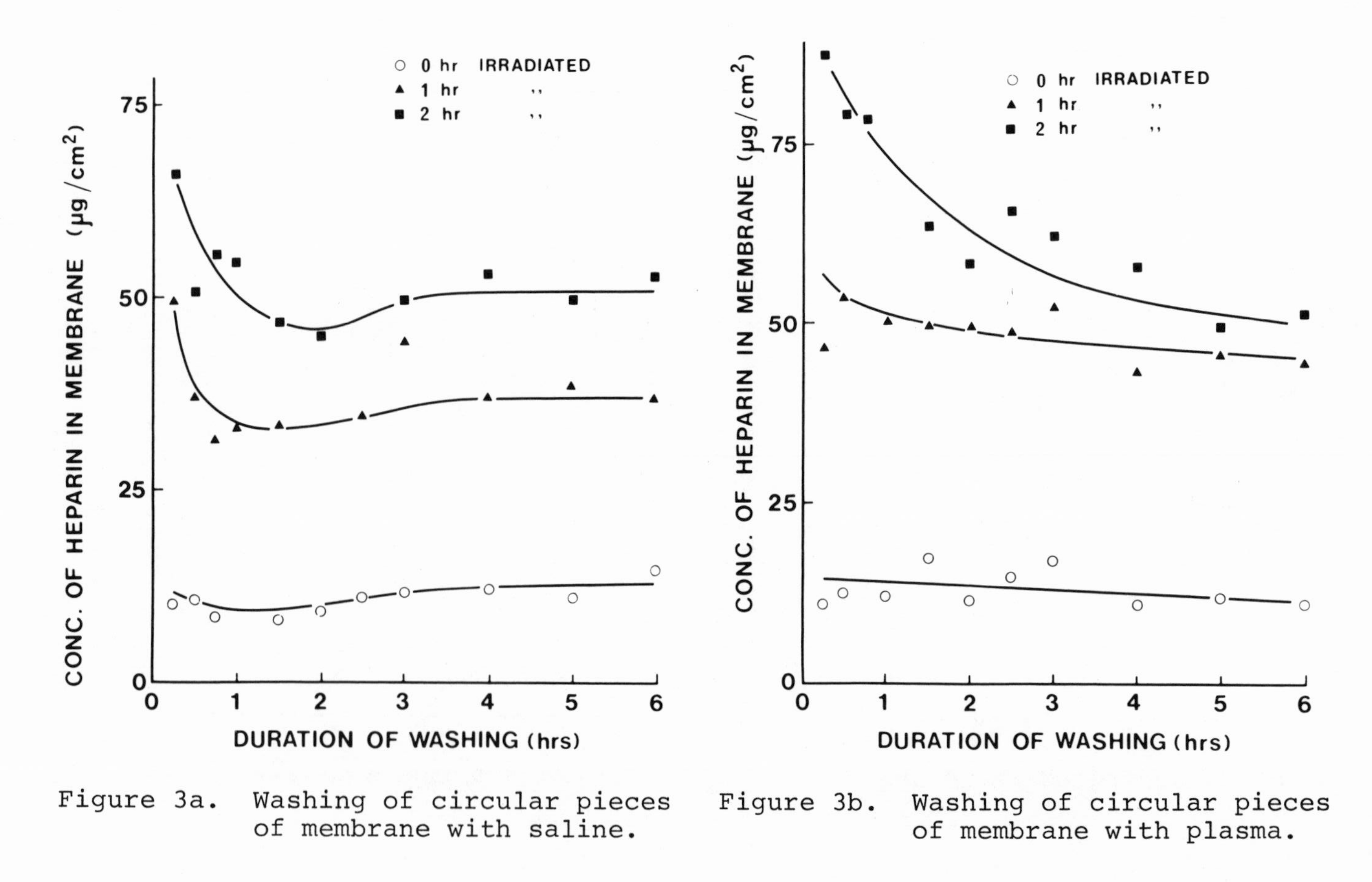

Figure 3a. Washing of circular pieces of membrane with saline.

Figure 3b. Washing of circular pieces of membrane with plasma.

TABLE II

Clotting Times

Surface	Clotting Times (Minutes)
Glass	5.0
DEAECA	6.5
Heparinized DEAECA, non-irradiated	12.5
Heparinized DEAECA, 1 hour irradiated	>60.0
Heparinized DEAECA, 2 hour irradiated	>60.0

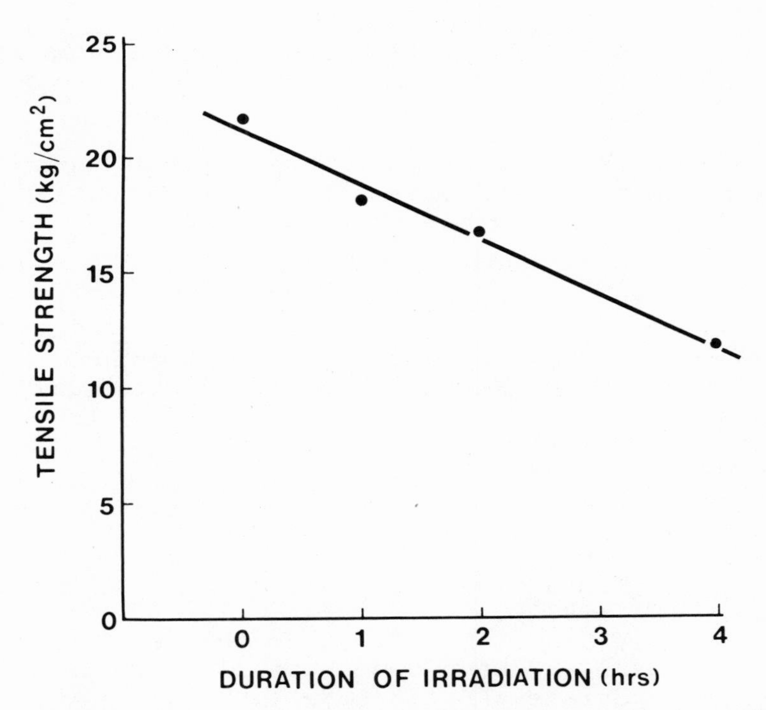

Figure 4 Effect of radiation on the tensile strength of membranes.

2 hour irradiated membranes did not clot even after 60 minutes.

After completing in-vitro experiments, in-vivo testing of heparanized test pieces were done. Details will be described in a later publication. A brief summary of the results is given here. Three different types of membrane materials were tested on dogs:

(i) Clots formed along the entire length of DEAECA which had not been exposed to heparin solution.

(ii) Clots of about 2 cm long was formed on the surface of DEAECA which had been soaked in heparin solution overnight and then soaked in a saline bath for 16 hours. Much bigger clots were formed on similar test pieces which had been soaked for two days in saline.

(iii) In the case of DEAECA which had been heparinized by the method presented in this report and then soaked in saline, the test pieces which had been soaked in saline for 16 or 40 hours showed either no clot or only very small white thrombi (<0.25 cm) on the surface.

From these results it would appear that heparinization by gamma radiation is promising as a new method for forming nonthrombogenic heparanized surfaces.

REFERENCES

1. Chang, T. M. S., "Perspectives in Biomedical Engineering" ed. R. M. Kenedi, Macmillian Press Ltd., London, 1973, ppl.
2. Gott, V. L., Whiffen, J. D., and Dutton, R. C., Science, 142:1297 (1963).
3. Gott, V. L., Whiffen, J. D., Koepke, D. E., Daggett, T. L., Boake, W. C., and Young, W. P., Trans. Amers. Soc. Artif, Int. Organs, 10:213 (1964).
4. Yen, S. P. S., and Rembaum, A., J. Biomed. Mater Res. Symp., 1:83 (1971).
5. Falb, R. D., Grode, G. A., Takahashi, M. T., and Leininger, R. I. Development of Blood-compatible Polymer Materials, PB 175 668, (1973).
6. Leininger, R. I., Epstein, M. M., Falb, R. D., and Grode,G. A., Trans. Amer. Soc. Artif, Intern. Organs, 12:151 (1966).

7. Merker, R. L., Elyash, L. J., Mayhew, S. H., and Wany, J. Y. C., Artificial Heart Program Conference, R. J. Hagyeli, ed., National Heart Institute, Washington, D. C., pp. 29 (1969).
8. Grode, G. A., Anderson, S. J., Grotta, H. M. and Falb, R. D., Trans. Amer. Soc. Artif, Intern. Organs, 15:1 (1969).
9. Grode, G. A., Falb, R. D., and Crowley, J.P., J. Biomed. Mater. Res. Sym. 3:77 (1972).
10. Eriksson, J. C., Gillberg, G. and Lagergren, H., J. Biomed. Material Res. 1:301 (1967).
11. Langergren, H. R., and Eriksson, J. C., Trans. Amer. Soc. Artif. Int. Organs, 17:10 (1971).
12. Merrill, E. W., Salzman, E. W., Wong, P. S. L., Ashford, T. P., Brown, A. H., and Austen, W. G., J. App. Physiology, 29:723 (1970).
13. Halpern, B. D., and Shibakawa, R., Advances in Chemistry Series 87:197 (1968).
14. Merrill, E. W., Salzman, E. W., Lipps, B. J. Jr., Gilliland, E. R., Austen, W. G., and Joison, J., Trans. Amer. Soc. Artif. Int. Organs, 12:139 (1966).
15. Martin, F. E., Shuey, H. F., and Saltonstall, C. W. Jr., J. Macromol. Sci. Chem. A4:635 (1970).
16. Chawla, A. S., Chang, T. M. S., Org. Coatings and Plastics Chem. Preprints, 33(2):379 (1973).
17. Malm, C. L., Tanghe, L. J., Laird, B. C., and Smith, G. D., J. Amer. Chem. Soc., 75:80 (1953).
18. Mahin, D. T., and Lofberg, R. T., Analytical Biochemistry, 16:500 (1966).
19. Chapiro, A., "Radiation Chemistry of Polymeric Systems," Interscience, New York, (1962), p. 533.

ACKNOWLEDGEMENTS

Thanks are due to Dr. Leon St. Pierre for the use of the Gammacell and Inston. The technical assistance of Miss E. Taroy is acknowledged. This work is supported by a grant from the Medical Research Council of Canada (MRC-SP-2) to TMSC.

RADIATION GRAFTED HYDROGELS ON SILICONE RUBBER AS NEW BIOMATERIALS*

Buddy D. Ratner and Allan S. Hoffman

Departments of Chemical Engineering
and Bioengineering
University of Washington
Seattle, Washington 98195

INTRODUCTION

Hydrogels have been considered as having great potential as biocompatible surfaces (1,2). The Annual Report of The Medical Devices Applications Program of the National Heart & Lung Institute (July 1, 1971 - June 30, 1972) reinforces this proposition in that hydrogels are often cited as being among the most promising of the materials tested. (Also see Ref. 3).

A primary difficulty in utilizing water-swollen hydrophilic cross-linked gels for prosthetic applications is their lack of mechanical strength. In order to form practical, mechanically tough materials with biologically compatible surfaces, hydrophilic monomers have been radiation graft polymerized to Silastic rubber (4, 10, 11). The monomers which have been used in this study are 2-hydroxyethyl methacrylate (HEMA) and N-vinyl-2-pyrrolidone (N-VP). Poly(HEMA) has been shown to have, in a number of situations, high degrees of biocompatibility (5,6). The polymer has the additional advantage of being relatively chemically stable (7). Poly(N-VP) has a long history of use as a blood extender. It is also a significantly more hydrophilic polymer than Poly(HEMA) based upon its high water solubility.

Radiation grafted hydrogels should also be ideal as

*This study was supported by Contract. (AT(45-1) 2225) with the U.S. A.E.C.

supports for the covalent attachment of proteins, enzymes, mucopolysaccharides, and heparin. This could produce surfaces of increased biocompatability or therapeutic devices for the treatment of pathologic conditions due to enzyme deficiencies. Initial work in this area has already been reported (8).

Based upon the potential uses cited for these grafted hydrogels certain properties would seem to be desirable. High water contents in the surface graft might produce a surface which had similar diffusive, mechanical, and interfacial adsorption properties to living tissue (particularly the endothelial lining of the blood vessels.) A high water content might also allow a larger fraction of the biomolecules bound to the surface to retain biological activity. Complete coverage of the support by the hydrogel should be necessary to insure that localized "non-compatible" regions do not exist on the surface.

A number of variables affecting the radiation graft polymerization of HEMA and N-VP have been explored. Results are used to formulate possible mechanisms for the graft polymerization and to decide which conditions might best be used to produce grafted polymers which have characteristics desirable for new biomaterials or enzyme supports.

EXPERIMENTAL METHODS

Dow Corning non-reinforced Silastic(R) films were washed for 5 minutes in 0.1% Ivory soap solution in an ultrasonic cleaner. They were subsequently given three five minute rinses in freshly changed distilled water in the ultrasonic cleaner. The films were stored in a 52% R.H. chamber.

HEMA monomer was obtained from Hydron Laboratories, Inc., New Brunswick, N. J. The monomer obtained had a diester level of 0.02% and a methacrylic acid level of 0.02%. N-VP monomer (Monomer Polymer Laboratories, Inc.) was purified by drying over Drierite and distilling at 42°C, 1 mm Hg. Ethylene glycol dimethacrylate (EGDMA) (Bordon Chemical Co.) was used as received. Distilled water and reagent grade methanol were used in all experiments.

The cleaned Silastic films were suspended in monomer solutions without removal of air, and irradiated at room temperature in a ca. 20,000 curie cobalt 60 source (courtesy of the College of Fisheries, University of Washington). The radiation dose used in all experiments was 0.25 mrad.

After grafting, the films were removed from the bulk external polymer. Adhering bulk polymer was cleaned by vigorously rubbing with a sponge soaked in acetone-water (50:50 v/v). The films were then thoroughly washed in two stirred acetone-water baths and four distilled water baths over a 24 hour period.

Water contents of the grafted films were measured by blotting the films between two sheets of Whatman #1 filter paper for ten seconds using a 285 g. weight to insure even, reproducible pressure and weighing immediately. Percent water in the graft was calculated using the relationship

$$\frac{W_w - W_d}{W_w - W_s} \times 100 = \% \ H_2O$$

where W_w is the weight of the wet, grafted film, W_d is the weight of the grafted film dried in a dessicator over anhydrone at 1 mm Hg for 24 hours, and W_s is the weight of the untreated Silastic Film. Reported water contents are the average of at least two determinations.

The weight of the graft per area (W_g) was obtained using the following relationship

$$W_g = \frac{W_d - W_s}{\text{initial film area}}$$

All IR spectra were taken on a Beckman IR-4 spectrophotometer. ATR spectra were recorded using a reflection angle of 45° and a KRS-5 crystal.

Elemental analyses were done by Galbraith Laboratories, Knoxville, Tenn. All analyses were performed in duplicate.

RESULTS AND DISCUSSION

A - Effect of Monomer Composition on the Grafted Copolymer

All grafting solutions contained 20% (by volume) monomer. The monomer consisted of mixtures of

HEMA, N-VP and (in some cases) ethylene glycol dimethacrylate in various proportions. The effect of monomer composition on the degree of graft for various solvent mixtures is shown in Figure 1. The variation in water contents for these grafted hydrogels is illustrated in Figure 2.

The composition of the surface graft can be investigated by observing its I.R. spectrum by the attenuated total reflectance (ATR) technique. The carbonyl absorption for the ester linkage in poly(HEMA) is found to occur at 5.70μ($\sim$1030 cm^{-1}) while the amide-type carbonyl in poly(N-VP) has an absorption at 5.92μ($\sim$1008 $cm.^{-1}$)(see Figure 3). The ratio of these two peaks is proportional to the fraction of each of the monomers in the graft. The proportionality is not necessarily linear, however. In order to convert the ratio of the heights of these two absorption bands to per cent of each monomer unit in the graft, a calibration curve was drawn. Actual graft compositions for the calibration curve were calculated by subjecting the films to elemental analysis (carbon, nitrogen, hydrogen, total silicon and inorganic silicon filler) and correlating the results with the ATR spectra of these films.

It should be noted that in the ATR spectrum of the pure N-VP grafted film (Figure 3-D) a residual "HEMA Peak" is seen at 5.70μ. In the spectrum of the pure HEMA grafted film (Figure 3-A) a residual "N-VP Peak" appears at 5.92μ. The reason why these residual peaks are observed is not clear. However they were taken into account in constructing the calibration curve.

Figure 4 shows the relationship between the weight per cent HEMA in the HEMA-N-VP graft (as calculated from ATR data) and the water content of the graft. As the per cent N-VP in the graft increases, the water content increases in a somewhat linear fashion. The points on this plot were taken from films examined in a number of different experiments over a five month period with, in many cases, different instruments settings. This might account for some of the scatter. Still, the general trend is quite apparent. Based upon this plot, an examination of Figure 2 indicates that in solvent systems containing only H_2O, N-VP enters into the graft much more rapidly than for methanolic solvent systems. This can be

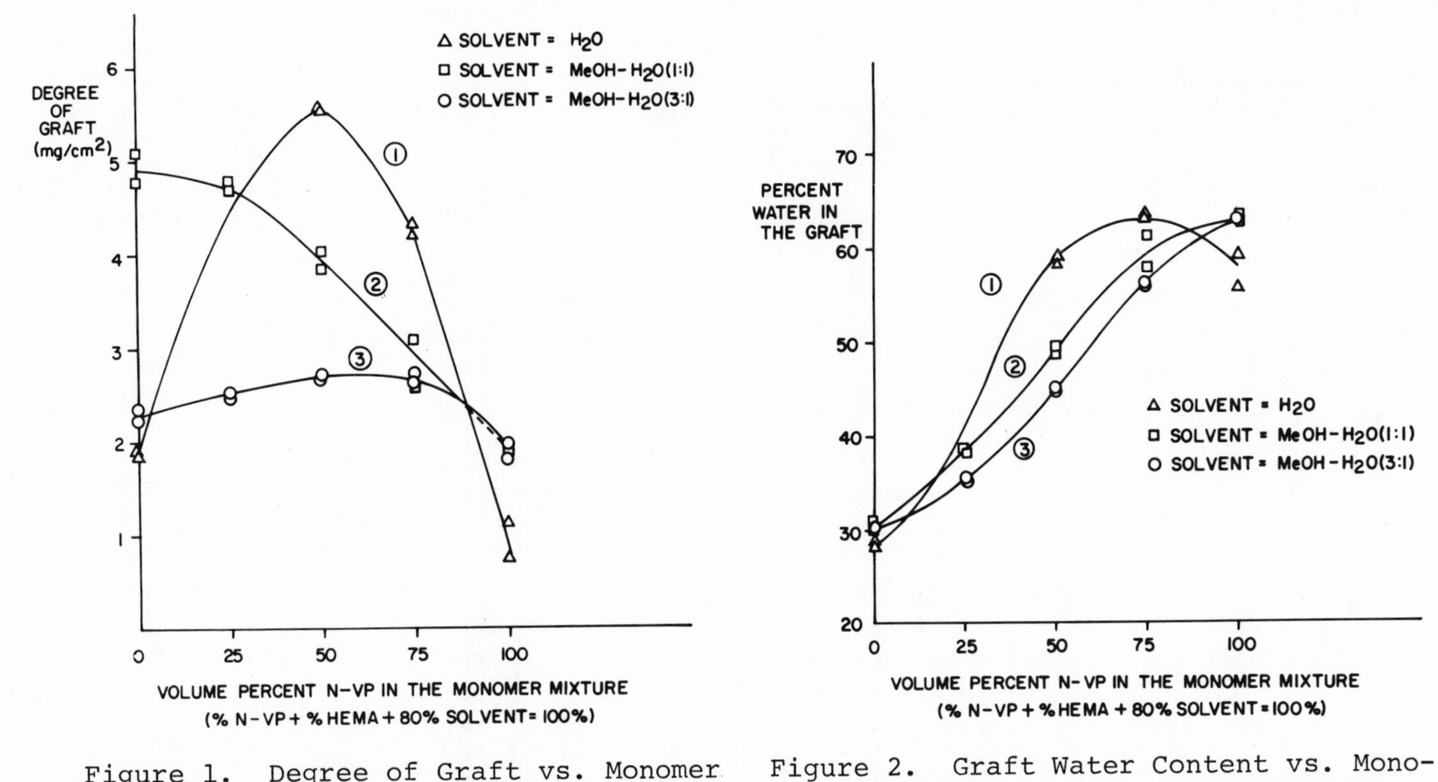

Figure 1. Degree of Graft vs. Monomer Composition for Grafted Hydrogels Prepared in Various Solvent Systems.

Figure 2. Graft Water Content vs. Monomer Composition for Grafted Hydrogels Prepared in Various Solvent Systems.

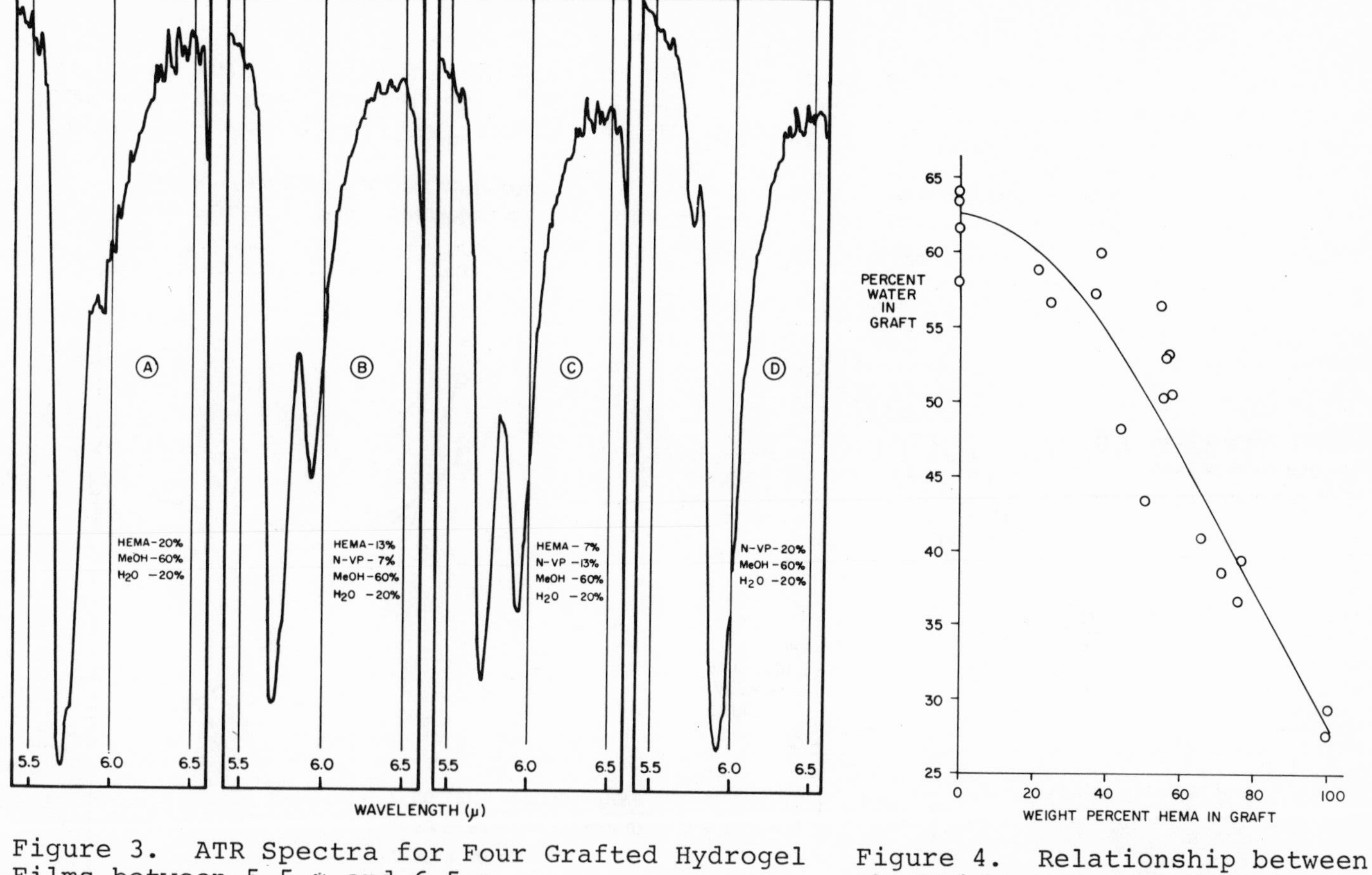

Figure 3. ATR Spectra for Four Grafted Hydrogel Films between 5.5 μ and 6.5 μ.

Figure 4. Relationship between the Weight Per Cent HEMA in HEMA-N-VP Grafted Hydrogels as Calculated from ATR Data and the Per Cent Water in the Graft.

explained by considering the degree of partitioning of N-VP monomer into the Silastic in the presence of various solvents. In systems containing mainly H_2O, N-VP may preferentially partition itself into the Silastic phase. However, in primarily methanolic systems the N-VP may be more soluble in the external solution. Measurements of partition coefficients are underway to determine the validity of this argument. The drop in graft water content for curve 1, Figure 2 at the 100% N-VP end is unexpected. We believe that N-VP monomer, when grafted in H_2O, penetrates the Silastic forming a homogeneous hydrogel-silicone rubber material. Evidence for this is based upon observations of grafted hydrogel films during hydrolysis with 2.5N sodium hydroxide at 100°C for 2 hours. For films grafted with only HEMA monomer, a surface graft can be peeled off after this treatment. The silicone rubber support does not appear affected. For N-VP grafted hydrogels, however, the entire film uniformly degrades, loses its physical strength, and ultimately becomes a sludge. It is assumed that N-VP enters into the silicone rubber matrix, separates the poly-(dimethyl siloxane) chains and becomes covalently bound to them. Upon base hydrolysis the N-VP side chains are hydrolyzed and the whole system falls apart. It is the presence of the N-VP inside the Silastic in close proximity to the hydrophobic poly-(dimethyl siloxane) chains which may produce the observed drop in water content.

The degree of grafting for different monomer compositions is seen to be highly solvent dependent (Figure 1). A marked increase in the degree of graft is evident for mixtures of HEMA and N-VP monomer over either monomer alone in systems containing 80% H_2O. It is interesting and probably significant that the maximum in this curve corresponds to the grafting solution composition which apparently represents the transition region between a gelled external polymer (at higher HEMA contents) and a viscous liquid (at higher N-VP contents). A very similar grafting behavior has been noted for the radiation grafting of styrene-acrylonitrile mixtures onto rayon in the presence of H_2O(9).

B - Effect of Solvent Composition on the Grafted Copolymer

By varying the volume per cent methanol in the solvent one can produce extremem changes in the grafting behavior for a given monomer composition (see Figures

5,6) (Also see Ref. 10). Methanol probably serves two functions in this process. In an H_2O solvent, monomer mixtures containing HEMA are polymerized in a precipitating medium. As soon as even a small amount of external polymer has formed, a "precipitate" gel is present to inhibit the diffusion of unreacted monomer to the growing chains at the surface of the film. In addition, the gel effect enhances the rate of polymerization in the external solution thus depleting further the monomer available for surface grafting. As the per cent methanol in the solvent is increased, however, the medium becomes a better solvent for poly(HEMA). Simple experiments show that poly(HEMA) is indeed considerably more soluble in methanol-H_2O mixtures than in either methanol or water. The gel, therefore, does not form at intermediate methanol concentrations and the moderately viscous external polymer solution still allows free monomer diffusion to the film surface. Also, the viscous solution inhibits chain-chain radical termination reactions. For this reason, the degree of graft should increase initially with increasing methanol. Methanol also seems to be an inhibitor of polymerization for this system. In a solvent consisting of 100% methanol, for all cases tried, there was no graft and no external polymer (as determined by precipitation). Therefore, as the methanol concentration is increased beyond a certain point, a decreasing degree of graft is seen. All curves in Figure 5 have been extrapolated to zero graft at 100% methanol. Although the experiments at 100% methanol were not done for all the systems shown in the figure, we feel this extrapolation is valid. Still, these grafting systems will be tried and the results reported at a later date.

The trends described in the preceding paragraph are particularly evident in systems containing 20% HEMA monomer. The pronounced maximum represents the point past which enough methanol is present to insure that there will be no gel formation (at 25% methanol this system is a thick, fully soluble paste). For the monomer systems represented in curves 2 and 3, Figure 5, the presence of N-VP in the copolymer greatly increases its solubility in aqueous solutions. With a polymerization solvent containing only 20% methanol these systems are fully liquid. In order to see if there is a maximum in either of these curves one would have to investigate methanol concentrations in the solvent between 0 and 20%. This has not yet been

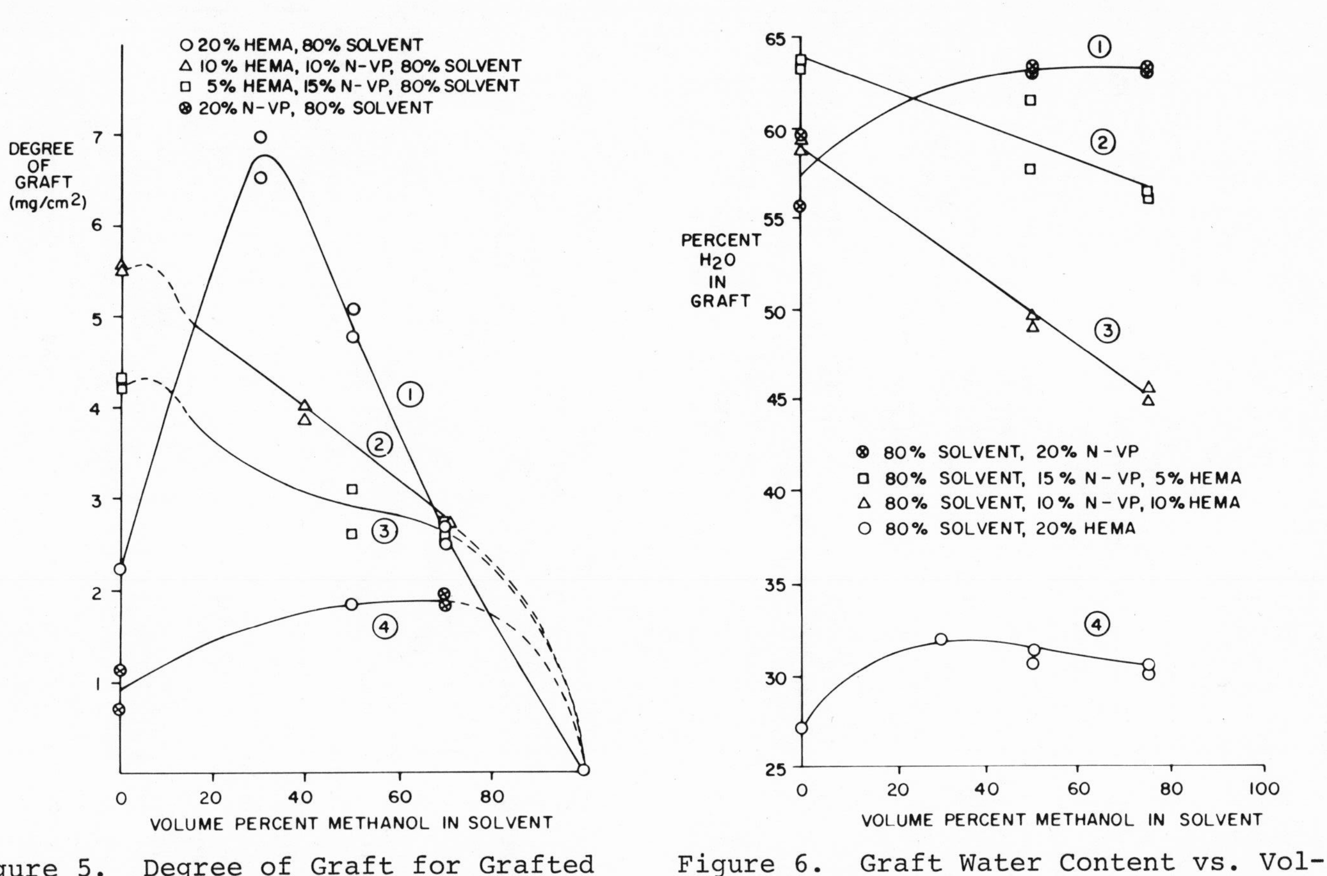

Figure 5. Degree of Graft for Grafted Hydrogels Prepared with Various Monomer Systems vs. Volume Per Cent Methanol in the H_2O-Methanol Solvent Mixture.

Figure 6. Graft Water Content vs. Volume Per Cent Methanol in the H_2O-Methanol Solvent Mixture for Various Monomer Systems.

done. However, the possibility of the existence of the maxima has been indicated on the plots by dashed lines.

The N-VP system (curve 4, Figure 5) does not, at any point, form a gel in the grafting solution with 1/4 mrad radiation dose. However, the viscosity of external solution after irradiation is extremely high for the system containing only H_2O as the solvent. The viscosity may be high enough to impede monomer diffusion.

The water content trends shown in Figure 6 can be readily explained by referring to the explanations given in Section A. Figure 6 was constructed by crossplotting the data in Figure 2.

C - Effect of Crosslinking Agent Concentration on the Grafted Copolymer

One of the primary, and most difficult to remove impurities in 2-hydroxyethyl methacrylate monomer is ethylene glycol dimethacrylate. Although the monomer used in these experiments had a very low diester content, it is possible that for commercial applications purification to such low diester levels would be impractical. Therefore, an understanding of the effect of the crosslinking agent in forming radiation grafted hydrogels is of value. Also, the addition of cross-linker has been found to be a successful method for controlling the degree of graft in certain of the systems discussed below.

The effects of added EGDMA on the degree of graft and the graft water content for various systems are shown in Figures 7 and 8. Grafted hydrogels formed in systems containing only HEMA and H_2O show decreasing graft with increasing crosslinking agent. Systems containing N-VP and a 60% MeOH-20% H_2O solvent, on the other hand, show increasing graft with increasing EGDMA. These effects can be explained by considering the point during the irradiation at which gelation occurs. For HEMA in H_2O, all homopolymers gelled. The higher the crosslink agent concentration, the earlier in the process at which the gelation will occur in the grafting solution. The gel hinders monomer diffusion to the film surface and encourages faster utilization of the remaining monomer so that the earlier it is formed,

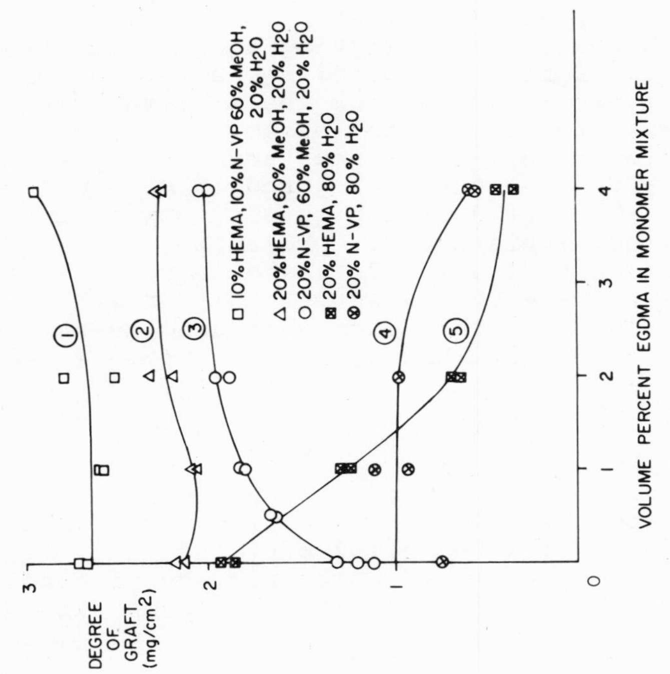

Figure 7. Degree of Graft vs. Volume Per Cent EGDMA in the Monomer for Grafted Hydrogels Prepared in Various Solvent Systems.

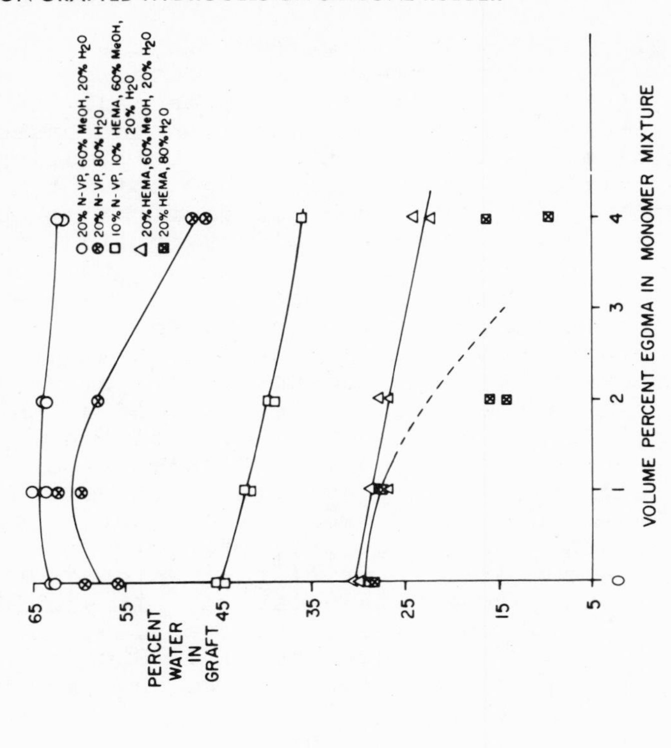

Figure 8. Graft Water Content vs. Volume Per Cent EGDMA in the Monomer Mixture for Grafted Hydrogels Prepared in Various Solvent Systems.

the lower the degree of graft. For N-VP in the methanolic solvent system there is no evidence of gelation until the 4% EGDMA level is approached. However the viscosity of the external solution increases with added crosslinking agent so that the termination reaction is inhibited. For N-VP in an H_2O solvent system, gelation with a 1/4 mrad. dose is first noted at the 2% diester level. The homopolymer is liquid, but of exceptionally high viscosity up to the two per cent level. It is possible that this homopolymer can hinder monomer diffusion, thus accounting for the relatively low grafts observed with this system. The other systems shown in Figure 7 do not fully gel until the 4% EGDMA level is reached. The gel formed at this point is extremely soft and of high liquid content. At the 2% diester level there is some insoluble slime evident in the external polymer solution. The small changes noted in the degree of graft as a function of added crosslink agent can be accounted for with arguments based upon a competition of effects which increase and decrease the graft as discussed above. It should be noted that the "dip" in curve 2, Figure 7, at the 1% EGDMA level is completely reproducible. Such a dip may also appear in curve 1 although we have not yet attempted to reproduce this.

In general, a decreasing water content in the graft is seen with increasing crosslinking agent concentration. This trend is as might be expected. The reason for the increase in water content for the N-VP-water systems between zero and 1% EGDMA is not clear. The effect is certainly related to similar trends seen in Figures 2 and 6. For grafted hydrogels prepared in HEMA-water mixtures, the degree of grafting is very low in the high EGDMA region. This usually presents great difficulties in measuring the very small quantity of water in the extremely thin graft. Thus, graft water contents as measured by the techniques used in this paper are considered to be somewhat unreliable in this region.

CONCLUSIONS

Silicone rubber materials with hydrophilic surfaces have been prepared by radiation grafting 2-hydroxyethyl methacrylate (HEMA) and N-vinyl-2 pyrrolidone (N-VP) monomers to Silastic rubber under a variety of conditions. The variables explored in this work were composition of the monomer mixture, composition of the grafting solvent, and per cent crosslinking agent in the monomer. All samples were irradiated with a 1/4 mrad.dose in a Co^{60} source. The degree of graft and the water content of the graft were measured. It was found that the degree of graft and graft water content could be varied over a wide range with only small changes in the grafting conditions. Particularly striking effects were (1) the pronounced maximum seen with HEMA monomer in the degree of graft as the per cent methanol in the methanol-water solvent mixture was varied, and (2) a synergistic increase in the degree of graft for 50:50 N-VP-HEMA monomer mixtures in an H_2O solvent systems. Based upon the results of these experiments, grafted hydrogels can now be prepared with grafts ranging continuously up to 6.5 mg./cm^2 and graft water contents ranging from 10% to 65%. The interaction of these different hydrogel surfaces with the natural environment (e.g. blood, tissue fluids) is being studied.

REFERENCES

1. O. Wichterle and D. Lim, Nature, 185, 117 (1960).
2. B. Levowitz, J. LaGuerre, W. Calem, F. Gould, J. Scherrer and H. Schoenfeld, Trans. Amer. Soc. Artif, Int. Organs, 14 82 (1968).
3. S. D. Bruck, Biomaterials, Medical Devices, & Artificial Organs, 1,79 (1973).
4. A. S. Hoffman and C. Harris, Polymer Preprints 13(2), 740 (1972).
5. A. Warren, F. Gould, R. Capulong, B. Glotfelty, S. Boley, W. Calem and S. Levowitz, Surgical Forum 14 183 (1968).
6. I. Michnevic and K. Kliment, J. Biomed. Mater. Res. 5,17 (1971).
7. S. Sevcik, J. Stamberg, P. Schmidt, J. Polymer Sci. Part C, No. 16,821 (1967).
8. A. S. Hoffman, G. Schmer, C. Harris, W. G. Kraft, Trans. Amer. Soc. Artif. Int. Organs, 18, 10 (1972).
9. W. H. Rapson and E. Kvasnicka, Tappi, 46,662 (1963).
10. A. S. Hoffman and W. G. Kraft, Polymer Preprints,13 723,(1972).
11. H. B. Lee, S. H. Shim, J. D. Andrade, Ibid., 729 (1972).

STRUCTURAL REQUIREMENTS FOR THE DEGRADATION OF CONDENSATION POLYMERS *IN VIVO*

P. Y. Wang and B. P. Arlitt

Institute of Biomedical Engineering

University of Toronto

Toronto M5S 1A4, Canada

INTRODUCTION

In the past two decades, much effort has been placed on the development of synthetic polymers degradable *in vivo* for surgical applicatons. A number of such polymers has been reported, and implants of polyglycolate have already gained clinical acceptance (1). However, the relationship between the macromolecular fragmentation of an implant *in vivo* (bio degradation), utilization or transfer of the breakdown products by neighbouring tissue (bioabsorption), and the accompanying histological responses has not been fully investigated. Simple homopolymers, such as the polyglycolate, are not very useful as model systems in providing the required information.

The hydrolysis of an ester can be regulated by the structure of the acid and the alcohol. (2) Four water soluble diacids with electron-withdrawing substituents that will facilitate ester hydrolysis by the inductive effect have been condensed with a linear and an alicyclic diol (Table I). The lengths of the chain in the diols are limited to about 6 carbon atoms, because higher diols, especially those with extensive branching, are less soluble in aqueous media, and are reported to be resistant to microbial degradation (3). The polyester polyols prepared from

TABLE I

DEGRADATION AND ABSORPTION TIMES FOR THE POLYURETHANES PREPARED FROM VARIOUS POLYESTER POLYOLS

Diacid Component	Diol Component			
	Cyclohexane Dimethanol		Hexane-1,6-diol	
	Degradation (days)	Absorption $t_{1/2}$(days)	Degrad. (days)	Absorp. $t_{1/2}$(days)
Perfluoroadipic Acid	14	80	14	60
2,2-Oxydiacetic Acid	21	120	45	170
2-Oxoglutaric Acid	28	155	not implanted	
Tartaric Acid	35	180	35	125
(Mean Deviation)	($\pm$3 days)	($\pm$8 days)	($\pm$3 days)	($\pm$8 days)

the selected components were thick syrups or waxes; they were made into coherent polyurethane films for implantation in experimental animals.

EXPERIMENTAL

The polyester polyols were prepared by the direct melt method. Usually 0.03 mole of a diacid was reacted with 0.0315 mole of a diol in a large test tube. The mixture was agitated by nitrogen bubbling through the melt. After heating the test tube at 110°C for 1.0 hr., the temperature was raised and maintained at 135°C for 0.5 hr. To the mixture was added 0.005 g of aluminum chloride, and the nitrogen pressure was lowered to 0.005 mm of mercury during the next hour. In the following 4 hr, the temperature was maintained at 200°C. Finally it was brought to 235°C for another hour. The polyester polyol produced was cooled under nitrogen, and stored tightly capped.

The polyurethanes were made by first heating 1 g of the polyester polyol in an oven-dried test tube. A quantity of 1.5 ml of anhydrous tetrahydrofuran was slowly stirred into the warm polymer until a clear solution was formed. To this solution an amount of 0.1 g of tolylene diisocyanate was added slowly with rapid stirring, and the reaction was continued for one day before film casting on a Teflon sheet. The sheet was placed under a tetrahydrofuran saturated atmosphere for 1 day. The polyurethane was further cured in an oven at 60°C for one day. Strips of 40 mm by 5 mm from a 0.3±0.1 mm thick polyurethane film were fastened to a 0.1 mm thick Teflon ribbon (Fig. 1). Three such implants were placed subcutaneously in each male Wistar rat (Fig. 2). At regular intervals the implants were recovered to assess the degradation time, absorption time, and the severity of tissue reaction (Fig. 3). The time required for the implanted polymer to become an incoherent mass was considered as the degradation time *in vivo*. The absorption time was very difficult to determine without the use of implants labelled with radioactive material. At the final stage of absorption *in vivo*, the polymer residue was practically indistinguishable from the interstitial fluid. However, it was found that by comparing the original dimension and the maximum size of the implant swollen *in vivo*, the time of disappearance for one-half of the polymer could be estimated. The absorption half-time ($t_{1/2}$) was, therefore, taken as from the time

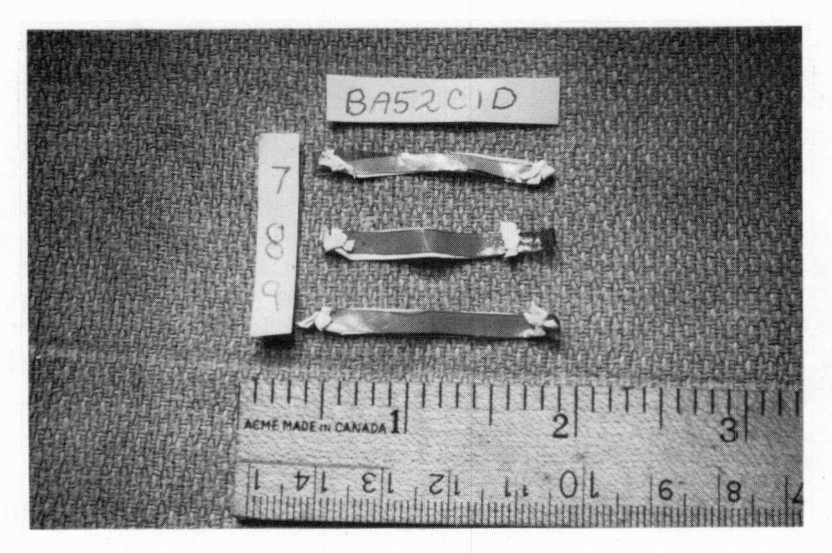

Fig. 1. Strips of diisocyanate-cured polyester fastened to Teflon ribbons used as controls

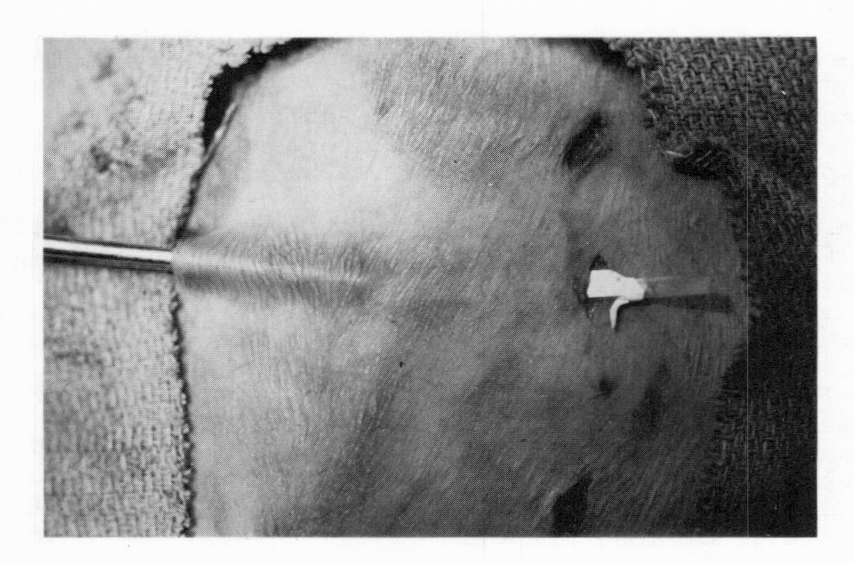

Fig. 2. An implant was being placed subcutaneously through stab wounds inflicted on the dorsal skin of an anesthetized Wistar rat.

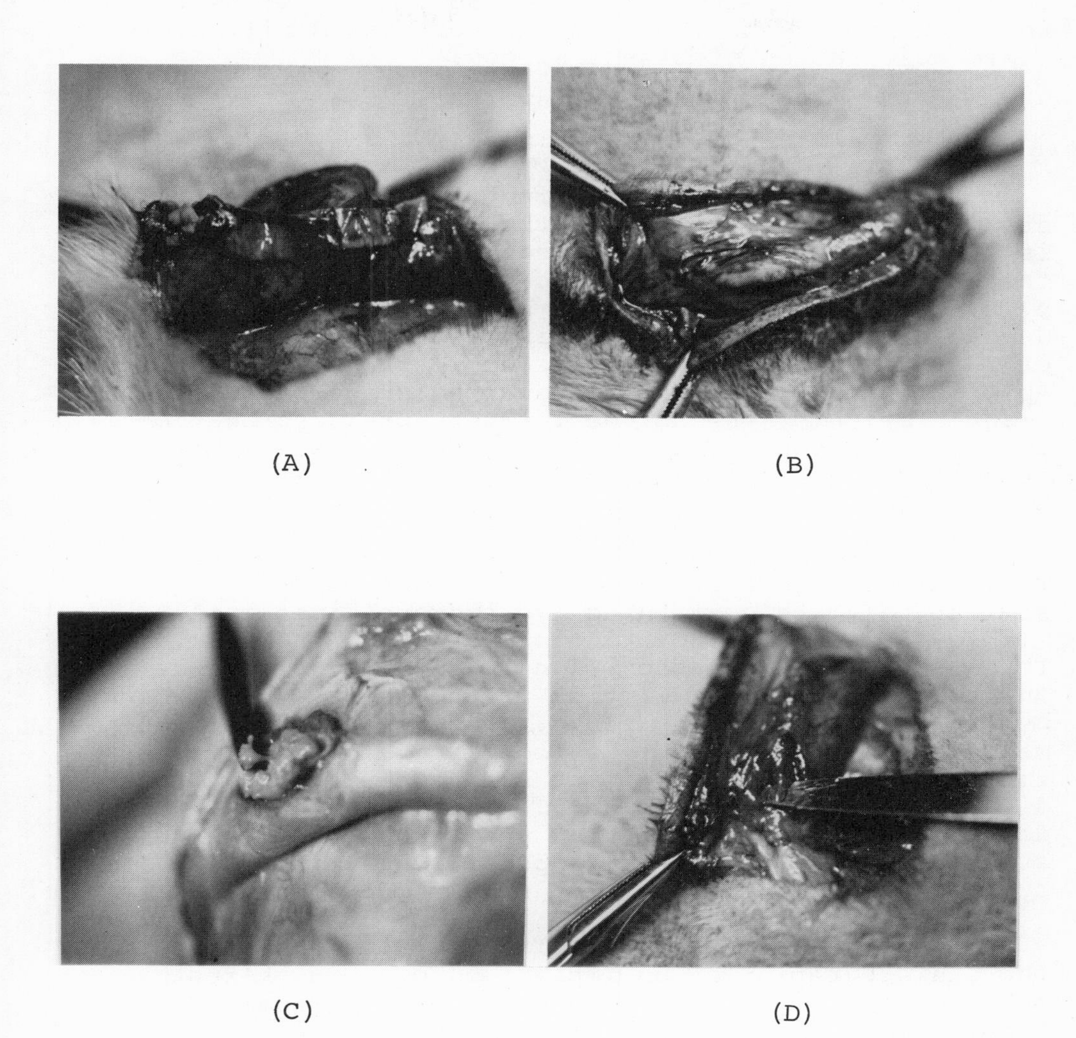

(A) (B)

(C) (D)

Fig. 3. The appearances of a polymer prepared from cyclohexane dimethanol and perfluoroadipic acid at various days after operation: (A) Essentially retained the original shape after 3 days; (B) Swollen after 8 days; (C) A syrupy mass after 60 days; (D) Absorption nearly completed after 120 days.

of degradation to zero strength until about one-half of the mass of the polymer had been absorbed. Gross tissue reaction around the implants was correlated with microscopic observation of tissue slides.

RESULTS AND DISCUSSION

Comparison of the data from the implantation studies (Table 1) shows that the absorption *in vivo* is directly proportional to the rate of degradation of the polyurethane. This is because the oligomers formed by random hydrolytic scission at the initial stage of degradation, though contributing little to the cohesion of the implant, were still insoluble in the body fluid. Absorption must occur at an advanced stage of molecular fragmentation of the polymer. Therefore, degradation *in vivo* is a continuous process, and the rate controlling step in the disappearance of an implant.

The polymers made by condensation of hexane-1, 6-diol with perfluoroadipic acid, and tartaric acid, respectively, were observed to be absorbed faster than similar polymers containing cyclohexane dimethanol (Table 1), although both diols are solube in water. This indicates that absorption *in vivo* is not completely due to solubility or dilution of the degradation products by the surrounding body fluid. Ingestion of the fragments by scavenging cells surrounding a disintegrating implant must also regulate to some extent the eventual disappearance of the implant. Microbial degration of detergents is known to favour unbranched hydrocarbon structures. (3) It is quite possible that phagocytosis is also a molecularly specific process.

The degradation time in Table 1 also shows that the disintegration of an implant into an incoherent mass *in vivo* is related to the electron withdrawing substituents alpha to the carboxyl functions in the diacid component. The strong inductive effect exerted by the fluorine atoms in perfluoroadipic acid has greatly facilitated the hydrolysis of ester linkages in the polymer which was completely degraded in two weeks. Minor structural variations in the diol did not appear to effect substantial changes in the rate of degradation *in vivo*.

The tissue reaction to the presence of most of these polymers was very mild, often being very difficult to distinguish from control implants (Fig. 4).

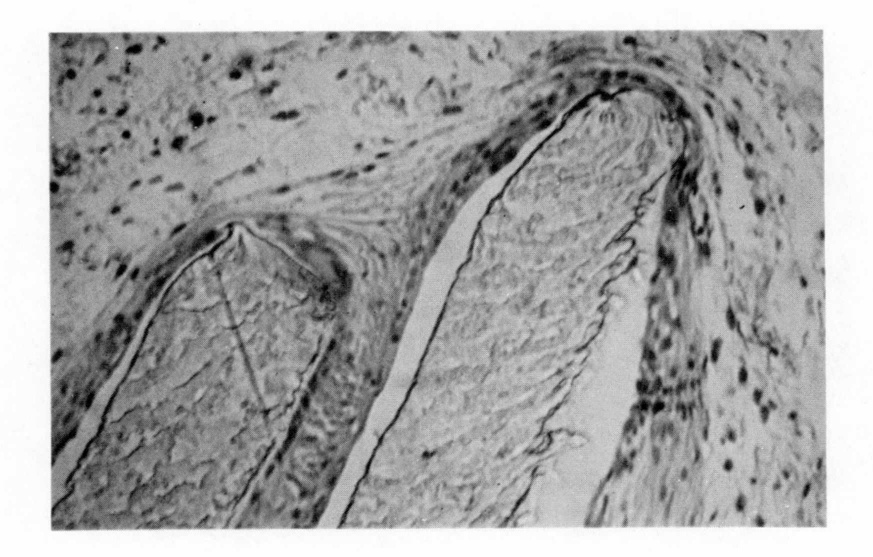

Fig. 4. Tissue response to a diisocyanate-cured polyester prepared from 2,2-oxydiacetic acid and cyclohexane dimethanol 4 days after operation. The strip on the left is the inert Teflon control. There is evidence of edema, but the cellular response to the degradable implant at right is very mild and comparable to the control (H & E stain, 200x).

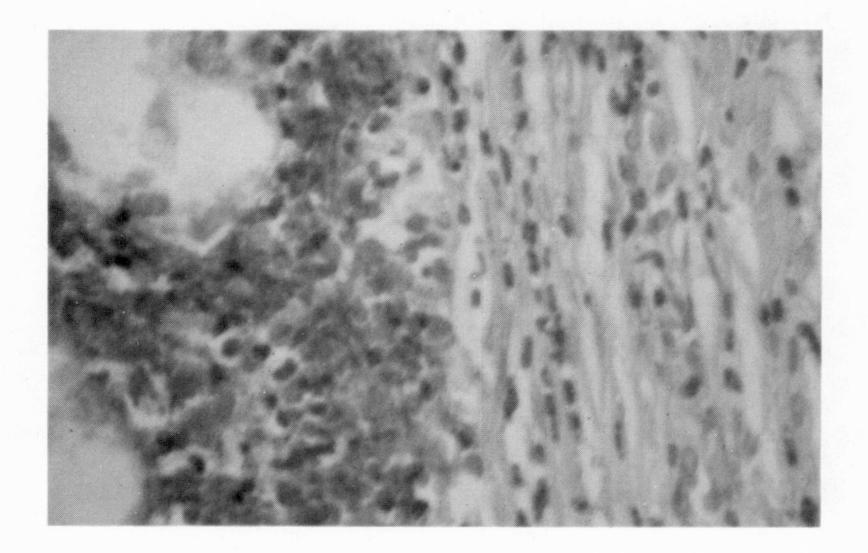

Fig. 5. Tissue response to a polymer prepared from perfluoroadipic acid and cyclohexane dimethanol 28 days after operation. There is a collection of polymorphonuclear exudate around the disintegrated implant at left. The connective tissue on the right also exhibits inflammatory response (H&E stain, 400x).

However, when an implant elicited an undesirable response, it was noted that this response was related to the rate at which the polymer was degraded and absorbed by the surrounding tissue. For example, in an extreme case, one polymer prepared from perfluoroadipic acid, glycolic acid, and glycerol had an absorption $t_{1/2}$ of 15 days. Accompanying the very rapid degradation and absorption was an intense inflammatory tissue response. Glycolic acid derived from the relatively slow degradation of its homopolymer has been shown to elicit only very mild response *in vivo* (4). The perfluoroadipic acid units in this polymer were also present in some of the other degradable implants (Table I), which degraded and absorbed much more slowly, and elicited no adverse tissue response. Therefore, this severe tissue reaction can only be attributed to the rapid degradation and absorption *in vivo*.

Implants containing perfluoroadipic acid and 2,2-oxydiacetic acid are better examples than the extreme case just discussed to illustrate the effect of degradation time and absorption $t_{1/2}$ on the neighbouring tissue. The polyurethane prepared from the polyester polyol made from the perfluoro acid and cyclohexane dimethanol degraded and absorbed 30% faster than the polymer consisted of the diacetic acid and cyclohexane dimethanol (Table I). Inflammatory tissue response was quite noticeable microscopically for the former polymer, while the latter implant elicited only mild irritation, indistinguishable as compared to the Teflon strip used as a control (Fig. 4,5).

Somewhat similar observations have been reported for the degradation of polyalkyl-2-cyanoacrylates [5]. The methyl ester derivative is degraded and absorbed rapidly with acute tissue reaction. The slower degrading butyl ester derivative is quite histocompatible.

Results obtained in this study strongly indicate that for a condensation polymer which is to be eliminated eventually *in vivo*, it must yield degradation products, preferably unbranched and with about six carbon atoms in order to achieve an adequate absorption $t_{1/2}$. However, the biocompatibility of such synthetic implants depends substantially on the degradation and absorption rate *in vivo*.

REFERENCES

1. Katz, A. R., Surg. Gynec. Obst., 131, 701 (1970)
2. Petai, S., "The Chemistry of Carboxylic Acids and Esters," p. 518, Wiley-Interscience, London (1969).
3. Patterson, S. J., Scott, E. C., and Tucker, K. B. C., J. Amer. Chem. Soc., 47, 37 (1970)
4. Herrman, J. B., Kelley, R. J., and Higgins, G. A., Arch. Surg. 100, 486 (1970).
5. Matsumoto, T., "Tissue Adhesives in Surgery," p. 38, Medical Examination Pub. Co., New York (1972).

DEGENERATION OF SILICONE RUBBER HEART VALVE POPPETS

M. S. Morgan

Carnegie-Mellon University
Chemistry Department
4400 Fifth Avenue
Pittsburgh, Pennsylvania 15213

INTRODUCTION

With the advent of the heart-lung machine, the correction of valvular heart disease by the use of prosthetic heart valves was introduced in 1961 (1). The prosthetic heart valve which has been used to replace the natural aortic, mitral or tricuspid valve has usually consisted of a silicone rubber ball incaged in a non-corroding type of chromium-nickel alloy cage (2). It is speculated that over 50,000 mitral and aortic valve prostheses have been supplied to date. Difficulties with the silicone rubber poppet were first reported in 1964 (3) and have since been confirmed in a number of publications (4-9). After varying periods of implantation, some of the poppets undergo marked changes in size and texture presumably due to absorption of blood lipids. The principal changes are development of a deep yellow color, swelling, loss of sphericity and elasticity, ultimate sticking and occasionally, fragmentation of the ball. These changes, collectively, have been termed "ball variance" in the literature. To circumvent this materials problem with silicone rubber, the past several years has seen the introduction of hollow metal or pyrolytic carbon balls and Teflon, carbon or polypropylene disks. Silicone rubber has also continued in use, especially since it is claimed that modification in the curing process of the rubber has greatly reduced ball variance (10, 11). In the interim the experience with poppets made of hard materials such as titanium or pyrolytic carbon has been somewhat less than satisfactory. A recent paper (11) on performance of materials in heart valve prostheses states that "the experience to date suggests that perhaps silicone rubber was abandoned prematurely by some valve manufacturers when in fact it was their curing and fabrication techniques which were at fault. In fact the evidence suggests that at the present time the improved-cure silicone rubber remains the material of choice for fabrication of the poppet in rigid prosthetic heart valves." Since only a limited percentage of poppets have failed thus far during their relatively short period of use, and there appears not to be a correlation of failure with length of time of implantation, this study considered the following alternatives: (1) those poppets that fail do so because of improper fabrication of the polymer, (2) failure is a natural consequence of silicone polymer swelling and will likely be an end result in all cases after varying lengths of time, or (3) there are abnormal constituents present in some patients' blood which are absorbed by the poppet. Most of the literature references refer to the material absorbed by the poppets as "lipid in character" and only recently have Carmen and Mutha (12) reported on specific chemical and spectroscopic data which led them to the conclusions that "the unknown is an oxidation product of polyunsaturated fatty acids." Our data, on the other hand, has led us to the conclusion that the lipid is a mixture of the oxidation products of cholesterol. Furthermore, we speculate that the oxidation takes place only after the cholesterol has been absorbed by the silicone rubber, and since it is finely dispersed in this matrix, it is autoxidized by molecular oxygen to which silicone rubber is known to be particularly permeable.

MATERIALS AND METHODS

Unused New Poppets

For control and reference purposes, we examined a series of 22 unused silicone rubber poppets derived from 3 manufacturers and include some poppets which are unfilled, two groups each with a different amount of $BaSO_4$ filler, and one group which represents varying cure times (Table I). The precise density measurements were made by applying Archimedes principle; that is determining the volume of the sample by weighing it

TABLE I. SAMPLES OF VARIOUS NEW SILICONE RUBBER POPPETS

Source of Specimens	No. Samples in Group	Density Range	% Extractable Av.	Swelling Ratio q
Dow-Corning Silastic Poppets	5	1.1352±.0029	2.40±.11	3.59±.26
Edwards Lab.				
Varying cure times	6	1.1396±.0013	2.49±.39	3.41±.24
Short cure + $BaSO_4$	2	1.1646	2.30	3.77
Surgitool, Inc. (All + $BaSO_4$)	9	1.2281±.0054	2.71±.21	3.61±.19

submersed in a liquid of known density. Generally this was done by weighing the sample submerged in water, however in determining the maximum swelling in neat chloroform, the ball was weighted and weighed while submerged in chloroform. The percentage of polymeric material extractable with a mixture of chloroform-methanol (2:1 v/v) at room temperature is given in column 2. After a specimen was extracted with the $CHCl_3/CH_3OH$ mixture, it was transferred to neat chloroform and the solvent changed daily until the specimen attained its maximum, equilibrium volume, as determined by application of Archimedes principle. The specimens were then transferred to a high-vacuum rack and degassed to constant weight. The ratio of maximum equilibrium volume observed in chloroform to the volume of the initial specimen is defined as the swelling ratio q. Since the degree of swelling observed at equilibrium in a good solvent invariably decreases with increasing degrees of cross-linking we determined this parameter as a means of evaluating the average degree of cross-linking.

<u>In-vivo Clinical Specimens</u> (Table II)

These samples were kindly supplied by Edwards Laboratories as was information on their implant time (in months). Of these six samples, only #37 contained some $BaSO_4$ to render it opaque to X-rays. Since in these six cases we were supplied with approximately one-fourth of the poppet, we have recorded the weight in column 3. The percentage of extractable with the $CHCl_3/CH_3OH$ (2:1 v/v) mixture was obtained both by determining the amount of non-volatile remaining after evaporating the extract and by loss in weight of the sample. These values agreed very closely and their average is given in column 5. The swelling ratio q* is given in column 4. This corrected q* was calculated from the following relationship:

$$q^* = q(\text{corr.}) = \frac{\text{Vol.}(\text{expanded in } CHCl_3)}{\text{Vol.}(\text{final})/.97 = \text{Vol. (orig.)}}$$

The calculated weight of siloxane in the extract (col. 8), is based on the assumption that 2.5 per cent of the weight of the original sample will be siloxane and will therefore appear in the total extract. Subtracting the mg of siloxane (col. 8)

TABLE II. EXPERIMENTAL DATA ON 6 CLINICAL SPECIMENS

Spec. No.	Impl. Time Mos.	Sample g	Swell. Ratio q*	Ext. %	mg	Col. Rec. %	Calculated Sil. mg	Calculated Lip mg	Chromat. Frac. mg.	2 %#	Sep. Frac. mg	7 %#
(1)	(2)	(3)	(4)	(5)	(6)	(7)	(8)	(9)	(10)	(11)	(12)	(13)
37	62	0.6554	3.74	14.6	97	88	15.5	80.0	10.8	70	54.8	69
38	53	0.9202	3.68	17.4	163	103	21.8	138.0	13.7	63	137.2	99
39	54	0.9805	3.80	19.4	194	94	23.2	168.0	16.7	72	142.6	85
40	55	0.7785	3.68	26.0	209	95	18.4	183.8	11.8	64	167.6	91
41	49	0.4844	3.56	20.5	100	96	11.4	87.7	7.2	63	64.0	73
42	40	0.4904	3.63	14.5	71	95	11.6	59.3	7.3	63	41.4	70

*The swelling ratio q on clinical specimens was obtained indirectly since the original volume was not available. However, from extraction data on the new poppets, it is known that after solvent extraction and complete degassing, the resulting $V_{(final)} = .97\ V_{(original)}$.

#These values are percentages of the calculated values in columns 8 and 9 respectively.

from the total wt of extract (col. 6) yielded the calculated weight of lipid (col. 9).

In order to achieve the separation of the composite lipid and siloxane mixture into its constituents, each of the extracts was subjected to an adsorption, liquid-solid chromatographic separation on silicic acid (Unisil), using a step-wise manual elution procedure. A preliminary column chromatographic run on the extract of a new poppet had indicated that a mixture of hexane-benzene (1:1 v/v) will elute 85% of the siloxane. A column chromatographic procedure was devised wherein each of the total extracts was successively eluted with the following sequence of elutants: (1) hexane; (2) hexane-benzene (1:1); (3) benzene; (4) benzene-chloroform (1:1); (5) chloroform; (6) chloroform-methanol (4:1) to the point where the yellow colored band was about to be eluted; (7) $CHCl_3/CH_3OH$ (4:1) in which the yellow colored band was obtained and (8) methanol. Each fraction was evaporated to dryness and a material balance performed. As noted in column 7, the column recovery was nearly quantitative in 5 of the 6 separations. Each fraction was dissolved in $CHCl_3$ to yield a 100 mg/ml solution which was used to obtain an infrared spectrum as a neat film between NaCl plates, and 100 µg of each of the 8 fractions from each clinical sample was used to prepare a thin-layer chromatogram. As a reference on each TLC plate, a preparation of human blood lipids was also spotted. The TLC plate depicted in Fig. 1 was developed with the solvent system 90 hexane/10 ether/1 acetic acid. This is the conventional developer for blood lipids and as may be noted in Fig. 1, all of the blood lipids except the phospholipids are moved from the origin. Since fraction 7 also remained at the origin, (although it is not phospholipid) an extensive effort was made to develop this fraction with a more polar solvent mixture (e.g. 100 chloroform/42 methanol/ 6 water which separates the phospholipids) but invariably we obtained a streaked plate without clearly discernable spots. Nevertheless, considerable analytical and spectroscopic information was obtained on concentrated, semi-purified fractions exemplified by fraction 7 obtained in the column chromatographic fractionation. Of the eight fractions collected, the weights and percentages of fractions 2 and 7 are given in columns 10 thru 13. Fraction 2 (hexane-benzene 1:1 v/v) is the major siloxane eluate and apparently represents 63 to 72% of the calculated amount of siloxane in the original extract.

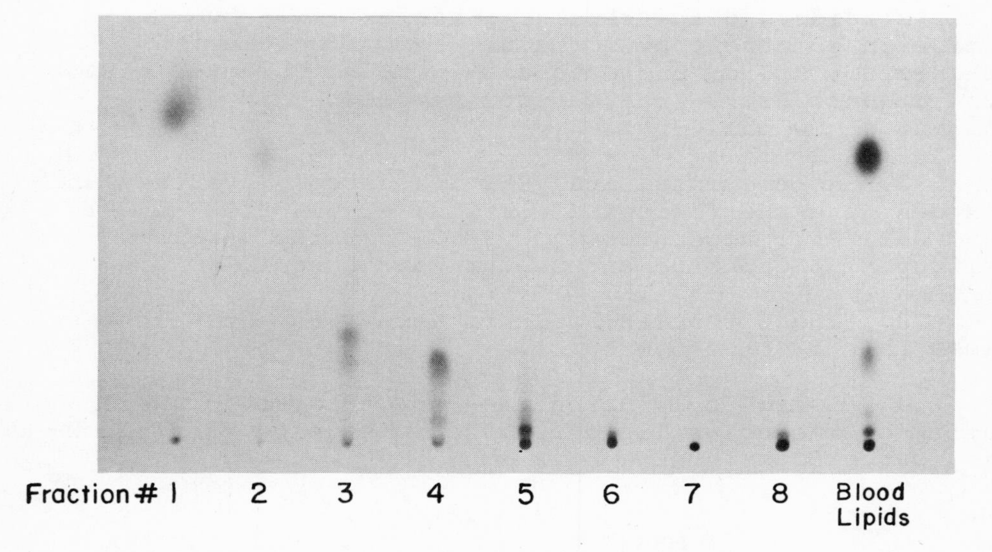

Fig. 1. Thin-layer chromatogram of the 8 fractions obtained by column chromatography of the extract of an in-vivo poppet compared with normal blood lipids.

Fraction 7, obtained with $CHCl_3/CH_3OH$ (4:1 v/v) elution is the polar lipid component (free of siloxane and neutral lipids) and represents 69 to 99% of the total amount of extractable lipid from the clincial specimen. It is the problem of purification and identification of fraction 7 which has received most of our attention.

EXPERIMENTAL RESULTS

1. A sample of a clinical specimen was wet-ashed with a mixture of concentrated sulfuric and fuming nitric acids and the resulting ash examined by emission spectroscopy. The amounts of calcium, magnesium, sodium and potassium were not greater than the amounts found in a control sample obtained by ashing a new ball.

2. An elemental analysis for carbon, hydrogen and oxygen gave the following results: C-67.89%; H-8.97% and 0-21.94%; accounting for 98.80% of the sample. Nitrogen, phosphorus and sulfur were absent.

3. The Lieberman-Burchard assay was equivalent to less than 5% of cholesterol.

4. A high resolution mass spectrographic examination of the polar lipid fraction disclosed the following parent ions:

Found m/e	Calculated Mass	Calculated Molecular Formula
400.3326	400.3341	$C_{27}H_{44}O_2$
384.3372	384.3392	$C_{27}H_{44}O$
382.3239	382.3235	$C_{27}H_{42}O$

In one in-vivo sample which had exuded some of the lipid, it was possible to examine a small sample which had undergone no separation nor chemical treatment whatsoever. The three major peaks observed were (a) most intense m/e = 366.3288; (b) 2nd most intense m/e = 382.3254 and (c) 3rd most intense m/e = 400.3349.

The largest parent ion detected, in either series of measurements was 400.333 which is equivalent to a molecular formula of $C_{27}H_{44}O_2$. Since this represents an increase of 14 mass units above the mol. wt. of cholesterol, it is suggestive of the oxidation of a methylene group to a carbonyl group. On the other hand, the oxidation of the 3-α hydroxyl to a ketone group followed by epoxidation of the 5,6-double bond would also account for a mol. wt. of 400.

5. The infrared spectra of the polar lipid fractions (fraction #7) of the six clinical specimens were essentially identical. Fig. 2-a is the spectrum of fraction 7 of specimen #39. It exhibits a strong C=O band at 1710 cm^{-1} (5.85μ), C-H band at a 2900 cm^{-1}, the 1460 and 1380 cm^{-1} bands due to CH_2- and CH_3- groups and a broad band ranging from 3600 to 2400 cm^{-1} indicative of hydrogen bonding. It is especially noteworthy that the infrared spectrum of fraction 7 (Fig. 2-a), of the exudate of a clinical specimen (Fig. 2-b) and the exudate of an in vitro test sample (Fig. 2-c) are very similar. The infrared spectrum of cholesterol (as a 1% soln in CCl_4, Fig. 2-d) is included for comparison. The hydroxyl (3650 cm^{-1} and C-O (1055 cm^{-1}) bands of cholesterol are missing in the other spectra, suggesting that the 3-β-ol has undergone some transformation.

6. The polar lipid fractions were dissolved in absolute ethanol and their ultraviolet absorption recorded on a Cary 14 spectrophotometer. The results are tabulated herewith:

Specimen #	λ max nm	$E^{1\%}_{1\ cm}$
37	227.5	77
38	222.5	45
40	230.0	55
41	222.5	60
42	227.5	71

No appreciable absorption occurred at wavelengths longer than the values cited, that is >230 nm indicating the absence of conjugation. If one assumes an average $E^{1\%}_{1cm}$ = 62 and an assumed mol. wt. of 400, then the absorptivity ε = ∿2500.

7. The nuclear magnetic resonance spectrum of one of the polar lipid fractions was kindly measured by Dr. R. F. Sprecher using a 250 MHz NMR Facility for Biomedical Studies made possible by a grant from NIH

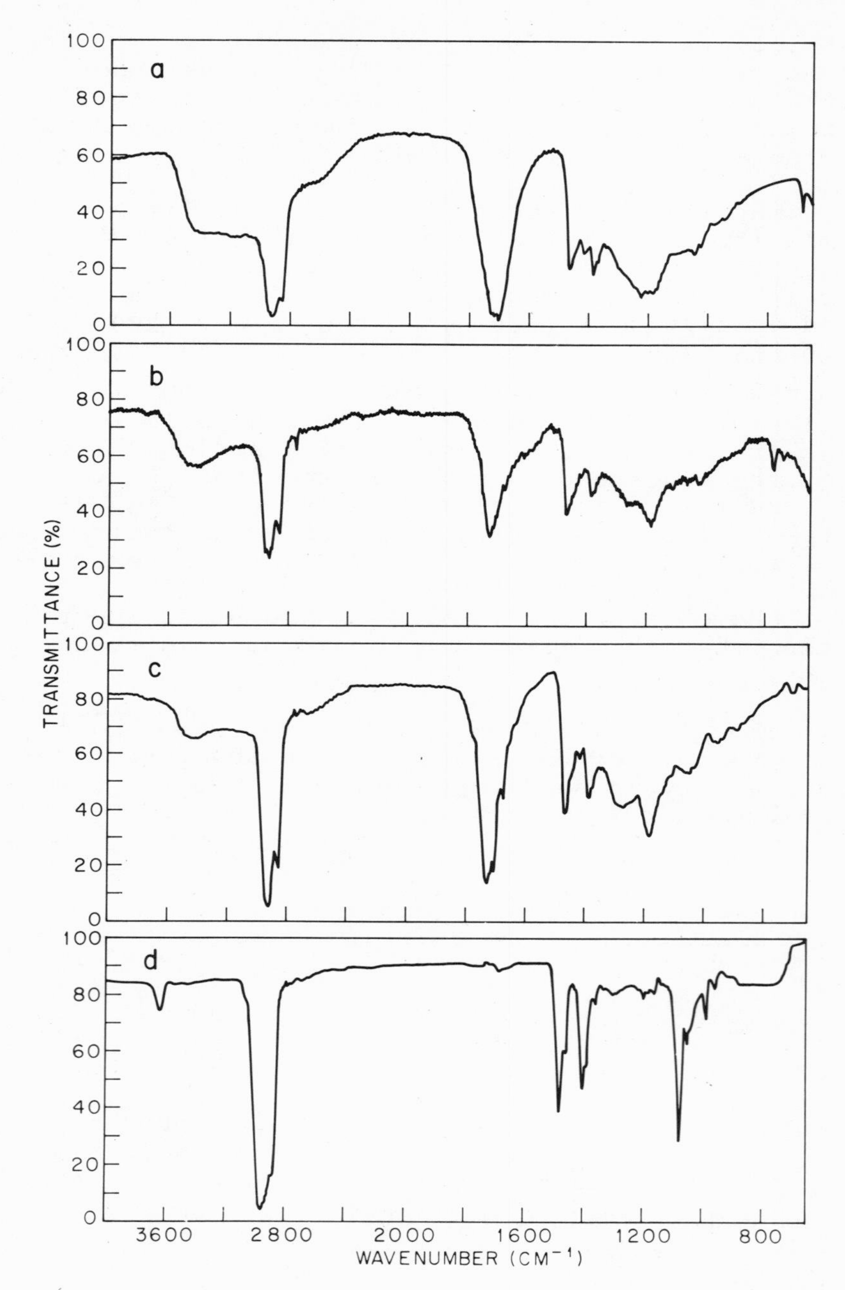

Fig. 2. Infrared spectra of (a) fraction 7 of specimen #39; (b) exudate of a clinical specimen; (c) exudate of the in vitro autoxidation experiment; (d) cholesterol (1% solution in CCl_4).

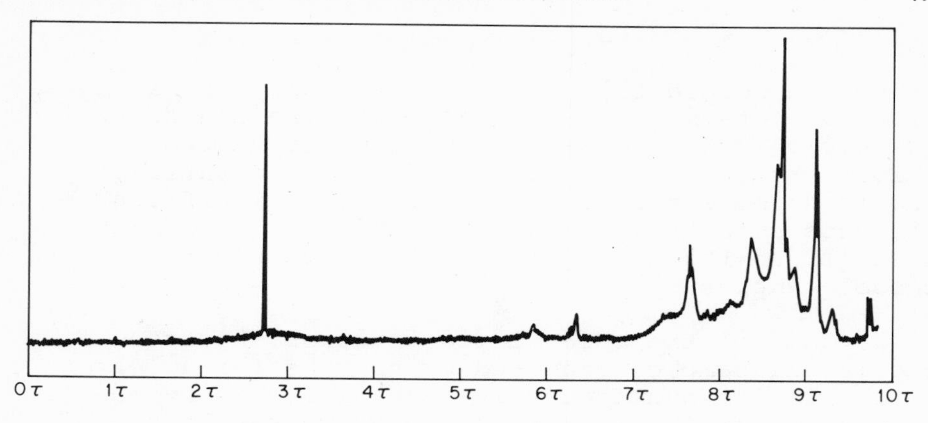

Fig. 3. NMR spectrum of the partially purified polar lipid isolated from an in-vivo specimen.

(NIH Grant RR-00292). The NMR spectrum is shown in Fig. 3 and is consistent with a steroidal configuration.

IN-VITRO AUTOXIDATION OF A CHOLESTEROL-IMPREGNATED POPPET

A 6g silicone rubber poppet was immersed in 100 ml of a 1% solution of cholesterol in chloroform and vibrated slowly on a shaker during 68 hours. Afterwards, the poppet was degassed on a high vacuum line to constant weight. A quarter-sphere was placed in a glass cylinder closed at the bottom by a fritted glass disc and the tube was kep immersed in a 37.5°C bath. Water-saturated compressed air was passed slowly thru the fritted disc over the specimen during several months. After an extended period of time it was noticed, inadvertently, that there were a number of colorless, viscous droplets on the surface of the specimen. These droplets (20 mg) were eluted with CCl_4 and an infrared spectrum recorded (Fig. 2-c). Surprisingly, the material displayed a complex, strong C=O band. A mass-spectrogram disclosed 4 parent m/e peaks at 368, 382, 384 and 400 respectively. A solution in absolute ethanol was examined for ultraviolet absorption. Although there was a small peak at 232.5 nm with an $E^{1\%}_{1cm}$= 57; no ultraviolet absorption was noted between 232 and 400 nm.

DISCUSSION OF RESULTS

1. In regard to the hypothesis that the failure of some poppets might be due to undercure and consequently a lower cross-link density; it must be concluded that in a comparison of swelling ratios, there is no evidence of a statistically significant difference between the average q of the new poppets (Table I) and the average q ($3.68 \pm .08$) for the six clinical samples.

2. All of the polar lipid fractions isolated from the clinical samples examined by us thus far appear to be very similar qualitatively, especially as evaluated by their infrared spectra. Although we have not yet succeeded in isolating an individual, chemically-pure substance, all of the physical and spectroscopic data reported herewith is consistent with a product characterized as a mixture of autoxidation products of cholesterol. This is confirmed by the in-vitro autoxidation experiment on the cholesterol impregnated sample.

3. On the basis of these findings we have dismissed the hypothesis that ball variance is due to the absorption of an abnormal component of blood.

4. These conclusions still leave unexplained the fact that ball variance occurs in a small percentage (2 to 5%) of cases and is not proportional to length of implant time. However the phenomenon of autoxidation is very sensitive to the presence of small amounts of oxidation-inhibitors and it may well be that these as yet unexplained results are related to the presence of as yet undetected promotors and/or inhibitors of autoxidation. If this speculation is correct then it would appear that ball variance could be prevented by incorporating a stable, immobilized antioxidant in the silicone rubber stock from which the poppet is fabricated, or impregnated into an already fabricated poppet.

ACKNOWLEDGMENTS

The author is indebted to Dr. T. H. Davies for suggesting this problem; to Dr. George Magovern, Allegheny General Hospital, Pittsburgh, Pa., and Dr. Samuel Koorajian of Edwards Laboratories for helpful discussions and making available most of the new and clinical materials for study; and to Mr. A. G. Wilshire for technical assistance during a part of this investigation. This research was supported in considerable measure by a Research Grant (HL-13075) from the National Heart and Lung Institute of the N.I.H.

REFERENCES

1. Starr, A., and Edwards, M. L., Mitral Replacement: The Shielded Ball Valve Prosthesis, J.Thorac. Cardiovasc.Surg., 42, 673 (1961).
2. Brewer, L. A., Editor-in-Chief, Prosthetic Heart Valves, Charles C. Thomas, Publisher, Springfield, Illinois, 1969.
3. Krosnick, A., Death Due to Migration of the Ball from an Aortic Valve Prostheses, J.A.M.A., 191, 1083 (1965).
4. Pierie, W. R., Hancock, W. D., Koorajian, S., and Starr, A., Materials and Heart Valve Prostheses, Annals New York Academy of Sciences, 146, 345 (1968).
5. Cooley, D. A., Bloodwell, R. D., Beall, A. C. Jr., Gill, S. S. and Hallman, G. L., Total Cardiac Valve Replacement Using SCDK-Cutter Prosthesis: Experience with 250 Consecutive Patients, Ann. Surg. 164, 428 (1966).
6. Hairston, P., Summerall, C. P., and Muller, W. H., Embolization of Silastic Ball from Starr-Edwards Prosthesis: Case Report and Comments, Ann.Surg. 166, 817 (1967).
7. Herr, R. H., et al: Diagnosis and Management of "Ball Variance" Following Aortic Valve Replacement, Circulation 36 (Supp. 2): 141 (1967).
8. Laforet, E. G., Death Due to Swelling of Ball Component of Aortic Ball-Valve Prosthesis, New Eng. J. Med. 276, 1025 (1967).
9. Starr, A., Pierie, W. R., Raible, D. A., Edwards, M. L., Siposs, G. G. and Hancock, W. D., Cardiac Valve Replacement; Experience with the Durability of Silicone Rubber, Supplement I to Circulation 33, 34, I-115 (1966).
10. Artificial Organs, A C&E News Feature, H. J. Sanders, Editor, C&E News, April 5, 1971, page 40.

11. Braunwald, N. S., Performance of Materials in Vascular Prosthetic Devices: Heart Valves, Bull.N.Y.Acad.Med., 48, 357 (1972).
12. Carmen, R. and Mutha, S. C., Lipid Absorption by Silicone Rubber Heart Valve Poppets - In Vivo and In-Vitro Results, J. Biomed. Mater. Res. 6, 327-346 (1972).

THE EFFECT OF IMPLANTATION ON THE PHYSICAL PROPERTIES OF SILICONE RUBBER

J. W. Swanson* and J. E. LeBeau**

*Rubber Research Dept., Dow Corning Corp.
**Toxicology Dept., Dow Chemical Co.
Midland, Michigan 48640

SUMMARY

Medical grade silicone rubber manufactured by Dow Corning Corporation was implanted subcutaneously in beagle dogs. At specified time intervals, the implants were removed at which time physical properties and weight changes were determined. A change in the physical properties was observed after six months implantation. Tensile strength decreased 7%, elongation 10% and the modulus at 200% extension increased 8%. After two years implantation, tensile strength had decreased 8%, elongation 15%, and 200% modulus increased 16%. After two weeks implantation, a 0.4% weight gain was seen, and after four weeks implantation the samples had gained 0.7%. However, after four weeks the weight gains stabilized and from four weeks to two years, they never exceeded 0.91%. The increase was due mainly to lipid absorption. No phospholipid or protein was detected. A chloroform/methanol extraction of the samples before implantation demonstrated 1.7% extractable silicone polymer. Samples not extracted before implantation, but extracted after implantation also demonstrated 1.7% extractable silicone polymer. This data suggests extractable silicone polymer does not leach from the medical grade silicone rubber during implantation.

The data also suggests the rubber is not being degraded to extractable silicone polymers during implantation. There were no significant differences in percent lipid up-take or the lipid profiles between samples which had extractable silicone polymer extracted before implantation and/or those which were not extracted before implantation.

It should be pointed out that the slight changes in physical properties measured in this study should not be great enough to alter the performance of the rubber in the subcutaneous environment.

INTRODUCTION

Silicone rubber has been widely used in medical applications requiring long-term implantation in the body. During its ten-year history in this area, silicone rubber has proven to be one of the most useful materials for implantation because of inertness and physical stability when in contact with body tissues and fluids.

However, in the use of silicone rubber for heart valve poppets, there have been reports that gradual deterioration or variance of the poppets can occur and lead to serious valve malfunction (1-4). It is believed that a major cause of variance is due to lipid accumulation from the blood into the silicone rubber (5-6). There have been several reports that all blood lipids are found in implanted silicone rubber poppets (3,5,7,8). However, a recent report by Chin and associates demonstrates that blood lipids are selectively absorbed by the poppets (6). No phospholipid or lipoprotein was found in the extracts of variant poppets they studied.

Raible and associates conducted engineering laboratory tests using variant and non-variant silicone rubber heart valve poppets in an effort to establish the degree of change in physical properties. (9) They concluded the up-take of lipids significantly reduces the tensile strength, ultimate elongation and abrasion resistance of silicone rubber. They also noted that extraction of the lipid materials from the variant poppets somewhat restores their physical properties, but they still remain appreciably lower than those of new production poppets.

It has also been suggested that the fracture of silicone rubber prosthetic finger joint implants is caused by the absorption of lipids (10). However, recently Meester and Swanson have suggested the absorption of lipids and the incidence of breakage of the implants are not related (11).

The purpose of this research was to study, under controlled experimental conditions, what effect implantation and the absorption of blood lipid materials have on the physical properties of silicone rubber.

EXPERIMENTAL

Materials

The same lot of medical grade silicone rubber manufactured by Dow Corning Corporation was used throughout the entire program. The base polymer in the rubber was a polydimethylsiloxane gum. The polymer had a small percentage of vinyl groups attached along the chain to enhance vulcanization. A high surface area silica filler was used to impart strength upon vulcanization. Vulcanization was carried out via a free radical mechanism using an organic peroxide at approximately 250°F. These slabs were post cured in an air-circulating oven for 24 hours at 300°F followed by 12 hours at 350°F. After post curing, dogbone shaped test specimens were cut out of the slabs for stress-strain analysis. These samples were sterilized by subjecting them to ethylene oxide at 130°F for eight hours. The samples were held for seven days to allow bleed-off of ethylene oxide before implanting.

Methodology

Three male and three female beagle dogs were used for the implantation studies. All dogs were approximately six years of age and had been acclimatized in our laboratory for six months. Purina Dog Chow and water were fed *ad libitum* throughout the study. Prior to surgery, each dog was fasted overnight. The next morning, the lower back and flank area on both sides of the spinal column were shaved and prepared for surgery. The dogs were anesthetized with 22mg/kg I.V. Surital (sodium thiamylal). For each silicone rubber implant, a 1 cm skin incision was made parallel to the spinal column. The subcutaneous fascia ventral to the skin incision was bluntly dissected producing a subcutaneous pocket large enough to hold the implant.

At predetermined times, the silicone rubber implants were removed from each dog. A skin incision and blunt dissection was used to isolate each implant. The fibrous capsule surrounding the implant was incised and the implant removed from the capsule. After isolation and removal, the implant was immediately weighed and physical properties determined within two hours.

Physical Property Determinations

Stress-strain curves were obtained on the dogbone shaped samples using an Instron tester. The Instron was equipped with a POGO extensometer to insure accurate elongation measurements. Pull rate was 2000% per minute. Fig. 1 shows the dogbone shaped specimen and a typical stress-strain curve obtained when it is tested. From curves such as this, ultimate tensile strength, elongation and 200% modulus were obtained.

In Vivo Studies Design

One objective of the research program was to determine the effect of implant time on the elastomer physical properties. Two male beagle dogs were used for this study. Each dog received dogbone shaped implants. The implants were removed in groups of five implants per group at 2,4,8,16,32 weeks and two year intervals for physical property testing. The two, four and eight week implants were all from one dog, and the 16 and 32 weeks and 2 year implants were from the second dog.

A six month implantation study was also carried out. Its purpose was threefold. First, to determine the effect of implantation on the physical properties of elastomer samples extracted prior to implantation. Second, to determine the amount and type of body materials absorbed into non-extracted and extracted silicone rubber samples. Third, to determine if any extractable silicone polymer had leached out of the implant.

One male and three female beagle dogs were used for this study. Each dog received five dogbone silicone rubber implants. The implantation time for all the

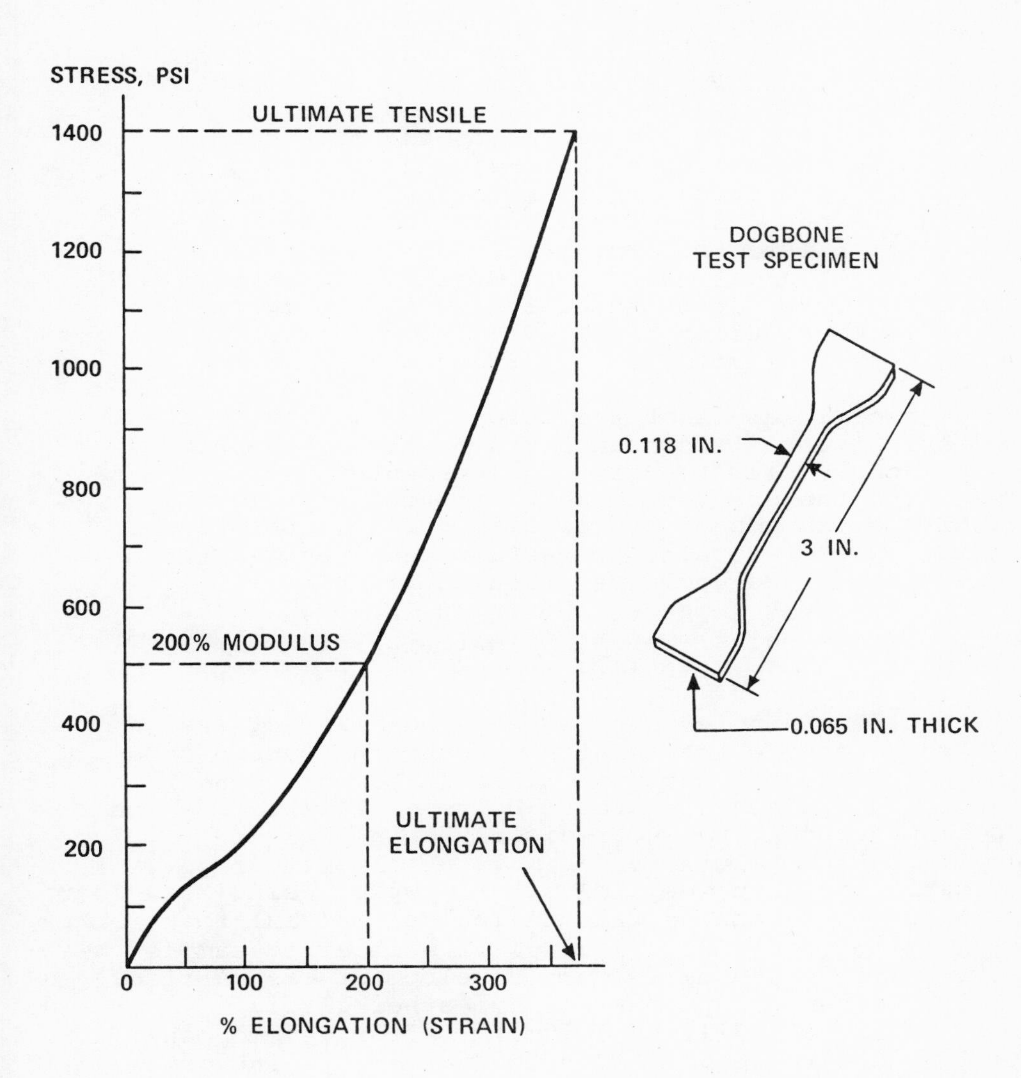

Fig. 1. Stress-Strain curve for medical grade silicone rubber

implants was six months. Before implantation, the specimens were divided into two groups (A and B) of ten per group. Implant specimens in Group A were extracted prior to implantation for 16 hours in a soxhlet extractor using a 2:1 by volume mixture of chloroform/methanol. The purpose of this extraction was to remove the majority of low molecular weight polydimethylsiloxane polymer which is not crosslinked into the rubber matrix. The specimens in Group B were implanted without being extracted.

After six months implantation, the specimens in both groups A and B were removed and their physical properties immediately determined. All of the samples were then extracted individually with 25 ml of chloroform/methanol (2:1 by volume) for 48 hours.

A two ml sample of the chloroform/methanol extract was removed for silicon analysis by atomic absorption spectroscopy. Atomic absorption measures only Si present in the extract. Therefore, in order to quantitate the amount of polydimethylsiloxane polymer present in the extract, it was assumed that all the silicon extracted would be in the form $[(CH_3)_2SiO]_x$ so a stoichiometric factor of 74/28 was used to calculate the actual weight of material extracted as silicone.

The remaining chloroform/methanol extract was used to determine the relative amount and type of body materials absorbed into the silicone rubber. The extract was placed in a round-bottomed flask, dried on a rotary evaporator, and redissolved in 2 ml of anhydrous diethyl ether. Thin layer chromatography was carried out on the 2 ml ether samples using 20 x 20 cm glass plates coated with silica gel (plain) to a thickness of 0.25 mm. Fifty µl of each ether sample was applied to the plate, and development was carried out in n-hexane/anhydrous diethyl ether/glacial acetic acid (85:15:1, by volume). After development, the plates were sprayed with Rhodamine 6G, ninhydrin, and molybdenum blue. The same plates were then sprayed with an even coat of 5N sulfuric acid and charred at 150°C.

RESULTS AND DISCUSSION

The Effect of Implant Time on Physical Properties

Table I summarizes the physical properties of the dogbone shaped samples after being implanted various times. The limits are the 95% confidence level on the mean of 5 samples at each time. The data on the eight groups of five samples, which include the control groups, were compared using F and T test statistics. Based on the results of these statistics, it can be stated with reasonable assurance that the physical properties of these samples did not change significantly with time through 16 weeks of implanting. However, at 24 and 32 weeks and two years, the physical properties are different when compared to samples stored in an environmental chamber six months at 37°C.

Table II summarizes the average weight increase of seven groups of five specimens removed at various times. The weight increases shown are those recorded three days after the samples were removed from the dogs. The weights immediately after removal from the dog averaged 0.10% higher than those shown. It was assumed the 0.10% weight loss was due to water evaporation from the samples. Note the weight increases leveled off after four weeks implant time. However, no change in physical properties was seen until 24 weeks.

Six Month Implantation Study

Results, summarized in Table III, show a change in physical properties of both extracted and non-extracted samples after six months implantation. Tensile and elongation values have decreased and the 200% modulus increased. Ultimate tensile strength and elongation depend upon the creation and propagation of flaws in the test specimens. Modulus is more dependent on the bulk structure of the material. For this reason, modulus values are more reproducible than tensile or elongation. Therefore, 200% modulus has been selected to demonstrate how physical properties are affected by sterilization, extraction, six months conditioning at 37°C and six months implantation. Results are shown in Fig. 2. Sterilization, six months conditioning at 37°C and implantation increases the 200% modulus of non-extracted medical grade silicone rubber. Extraction of the rubber does not change its 200%

TABLE I

PHYSICAL PROPERTIES OF MEDICAL GRADE SILICONE RUBBER AS A FUNCTION OF IMPLANT TIME

Time	Tensile Psi	% Elongation	200% Modulus Psi
0 Wks. (Control)	1388±33	377±20	506±10
2 Wks	1488±32	384±5	530±20
4 Wks	1320±61	359±3.3	539±29
8 Wks	1323±68	360±28	510±23
16 Wks	1352±65	373±9	518±7
24 Wks	1303±72	359±13	567±23
32 Wks	1294±13	354±6	559±21
Control (6 Mo. at 37°C)	1392±41	392±12	516±15

TABLE II

PERCENT WEIGHT INCREASE AS A FUNCTION OF IMPLANT TIME

Time	% Wt. Increase
2 Weeks	+ .40±0.08
4 Weeks	+ .75±0.20
8 Weeks	+ .70±0.20
16 Weeks	+ .83±0.15
24 Weeks	+ .91±0.23
32 Weeks	+ .70±0.14

TABLE III

PHYSICAL PROPERTIES OF MEDICAL GRADE SILICONE RUBBER AFTER 6 MONTHS SUBCUTANEOUS IMPLANTATION IN DOGS

Ref.*	Tensile Psi	% Elongation	200% Modulus Psi
N.E.	1392±41	392±12	516±15
E	1339±19	373±9	490±10
N.E.I.	1303±72	369±13	567±23
E.I.	1284±78	342±17	564±33

*N.E. -- Not extracted control group (5 specimens); 6 months at 37°C

E -- Extracted control group (5 specimens): 6 months at 37°C

N.E.I.-- Not extracted samples (10 specimens) that were implanted 6 months

E.I. -- Extracted samples (10 specimens) that were implanted 6 months

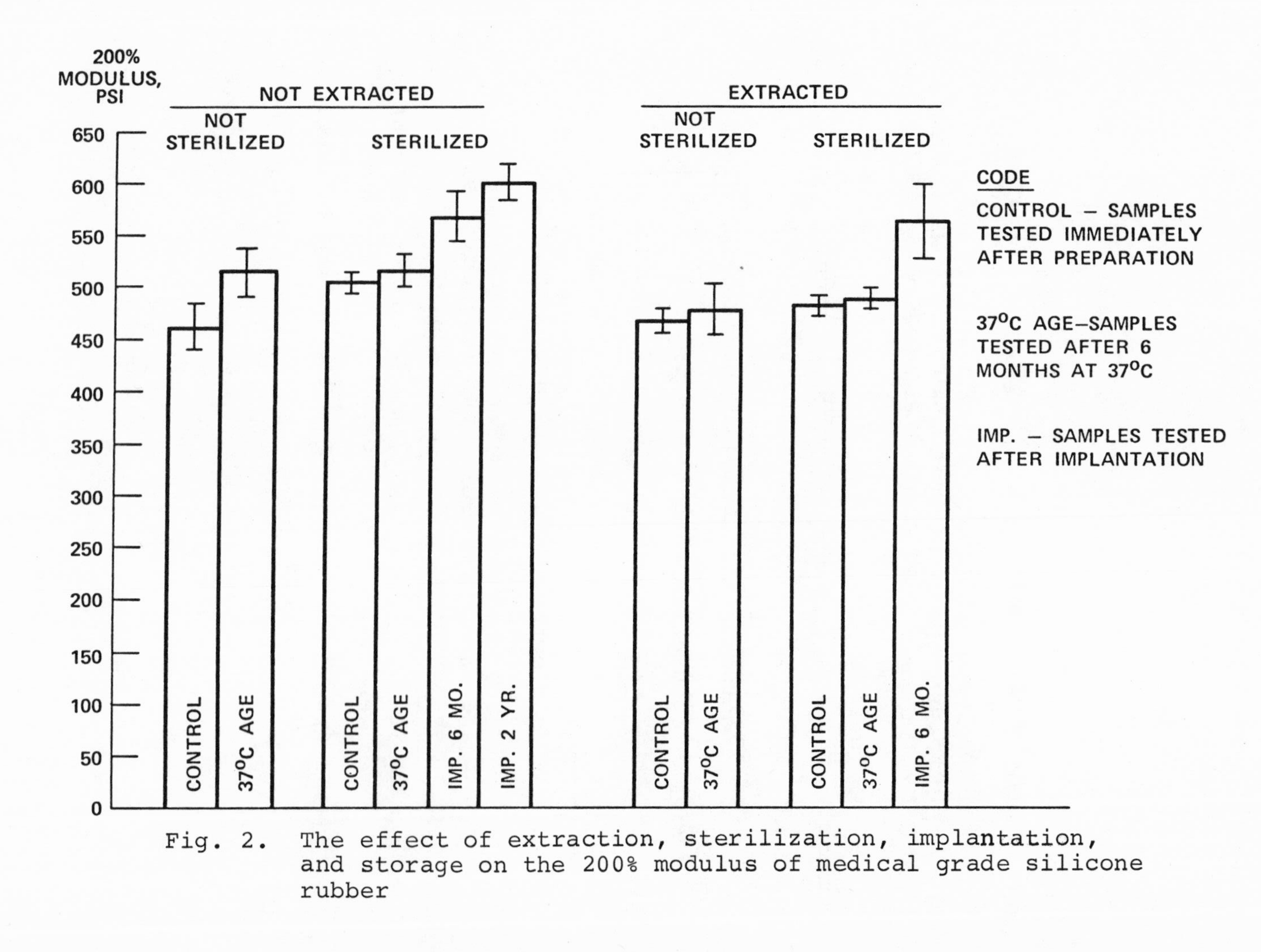

Fig. 2. The effect of extraction, sterilization, implantation, and storage on the 200% modulus of medical grade silicone rubber

modulus. Furthermore, once the rubber has been extracted, neither sterilization nor six months conditioning at 37°C seem to increase the 200% modulus. However, the 200% modulus of the extracted and non-extracted specimens are the same after six months implantation, but higher than the control specimens.

Table IV shows the percent weight increase, extractable silicone polymer, and extractable lipid after six months and two years subcutaneous implantation of silicone rubber in dogs. The extractable silicone polymer after implantation for previously non-extracted samples was 1.75% of the specimen weight. There was no significant difference between this value and the percent extractable silicone polymer in the samples extracted before implantation. Therefore, this data strongly suggests that the extractable silicone polymer is not leaching from the silicone rubber during implantation.

The extractable silicone polymer in the specimens extracted before implantation was 1.72%. After six months implantation, these same specimens exhibited 0.03% extractable silicone polymer. The value obtained when non-implanted specimens are extracted a second time is also 0.03%. This data suggests that the silicone rubber is not degraded to extractable silicone polymer during implantation.

All samples increased in weight during the implantation period. The average weight increase was 0.72%. There was no significant difference in weight increase between the samples extracted and the samples not extracted before implantation. Therefore, the presence of 1.7% uncured silicone polymer does not seem to influence the up-take of biological materials by silicone rubber.

The percent lipid extracted did account for a significant amount of the weight increase of the implanted samples. The relative proportion of lipids found is presented in Table V. Cholesterol and triglycerides were found in highest proportions. There was a small component left at the origin of the thin layer chromatogram which was assayed for protein and phosphorous. However, none was detected. This agrees with the results reported by Chin and associates in reference 6. The lipid profiles of implanted samples

TABLE IV

EXTRACTION ANALYSIS OF MEDICAL GRADE SILICONE RUBBER AFTER 6 MONTH IMPLANTATION

	Wt. Increase	Extractable silicone Polymer before implantation	Extractable silicone Polymer after implantation	Lipid Extracted
Group A (Extracted before Implantation)				
Dog 1 Female	.67±.19	1.70±.04	.03±.008	.52±.16
Dog 2 Male	.60±.07	1.74±.06	.03±.004	.52±.05
Group B (Extracted after Implantation)				
Dog 3 Female	.45±.09		1.80±.09	.40±.07
Dog 4 Female	.91±.23		1.70±.20	.91±.27

TABLE V

RELATIVE LIPID PROPORTIONS IN MEDICAL GRADE SILICONE RUBBER AFTER SUBCUTANEOUS IMPLANTATION IN DOGS FOR 6 MTHS

	Sterol Esters	Triglycerides	Free Fatty Acid	Cholesterol	Partial Glycerides
Group A (Extracted before implantation)					
Dog 1 Female	++	+++	+	+++	+
Dog 2 Male	++	+++	+	+++	++
Group B (Not extracted before implantation)					
Dog 3 Female	+	+++	+	+++	++
Dog 4 Female	++	+++	+	+++	+

which were extracted before implantation, were not significantly different from those which had not been extracted before implantation. This again indicates that the presence of 1.7% uncured polymer does not influence the up-take of lipids.

CONCLUSIONS

Six months subcutaneous implantation in dogs changed the physical properties of medical grade silicone rubber. Tensile strength decreased 7%, ultimate elongation decreased 10%, and the 200% modulus increased 8%. Two year implant studies demonstrate a further change in properties. Tensile strength decreased 8%, elongation 15%, and 200% modulus increased 16%. These changes in physical properties are slight and should not alter the performance of the rubber in the subcutaneous environment. Weight gain data demonstrate sample weights stabilize after four weeks implantation. However, changes in physical properties were not observed until six months. Extraction experiments demonstrated that the presence of extractable silicone polymer did not influence lipid up-take or physical property changes. The data also demonstrates that no extractable silicone polymer leaches into the body during the two year implantation period.

ACKNOWLEDGMENTS

The authors wish to thank Mr. Ben Franklin and Mrs. Chris Schmidt for their assistance on this project. Their efforts are sincerely appreciated.

REFERENCES

1. Starr, A.; Pierie, W. R.; Raible, D. A.; Edwards, M. L.; Siposs, G. G.; Hancock, W. D.: Cardiac Valve Replacement: Experience with the Durability of Silicone Rubber. In Cardiovascular Surgery 1965, edited by F. A. Simeone. American Heart Association Monograph 13. Supplements to Circulation 33 and 34 (Suppl I): I-115, 1956.
2. Laforet, E. G.: Death Due to Swelling of Ball Component of Aortic Ball-Valve Prosthesis. New Eng J Med 276: 1025, 1967.

3. Pierie, W. R.; Hancock, W. D.; Koorajian, S.; Starr, A.: Materials and Heart Valve Prostheses. Ann NY Acad Sci 146: 345, 1968.

4. Roberts, W. C.; Morrow, A. G.: Fatal Degeneration of the Silicone Rubber Ball of the Starr-Edwards Prosthetic Aortic Valve. Amer J Cardiol 22: 614, 1968.

5. McHenry, M. M.; Smeloff, E. A.; Fong, W. Y.; Miller, G. E. Jr.; Ryan, P. M.: Critical Obstruction of Prosthetic Heart Valves Due to Lipid Absorption by Silastic . J. Thorac Cardiovasc Surg 59: 413, 1970.

6. Chin, H. P.; Harrison, E. C.; Blankenhorn, D. H.; and Moacanin, J.: Lipids in Silicone Rubber Valve Prostheses After Human Implantation. Circulation 43 and 44 (Suppl I): 51-56 (1971).

7. Carmen, R.; Kahn, P.: Test in vitro of Silicone Rubber Heart-Valve Poppets for Lipid Adsorption. J. Ass Advance Med Instr 3: 14, 1969.

8. Noiret, R.; Penther, P.; Bensaid, J.; Beaument, J. L.; Lenegre, J.: L'infiltration Lipidique des Prostheses Valvulaires de Starr. J. Atheroscler Res 8: 975, 1968.

9. Raible, D. A.; Keller, P. P.; Pierce, W. R.; and Koorajian, S.: Elastomers for use in Heart Valves. Rubber Technol. 39: 1276-1286, 1966.

10. Homsy, C. A.: Biocompatibility in Selection of Materials for Implantation. J. Biomed. Mater Res., 4 (3) 341-56, 1970.

11. Meester, W. D.; and Swanson, A. B.; In vivo Testing of Silicone Rubber Joint Implants for Lipid Absorption. J. Biomed. Mater Res., 6: 193-199 (1972).

CORROSION STUDIES ON ANODE MATERIALS FOR IMPLANTABLE POWER SOURCES

Hatim M. Carim, Richard B. Beard,
Joseph F. DeRosa, and Steven E. Dubin
Drexel University
Biomedical Eng. & Science Program
Philadelphia, Pennsylvania

INTRODUCTION

Many thousands of patients have received successfully implanted pacemakers and live useful lives. Recent advances in technology have led to more dependable electrodes and generators so that the average functional life of a pacemaker has increased from six to eighteen months with some as high as thirty-six months (1,2). Ideally an implantable power source should last for ten years, eliminating the necessity of replacing power sources (3). There are a variety of approaches which have been pursued in an attempt to prolong the lifetime of internal power sources (4). The implantable hybrid cell as a long term power source for cardiac pacemakers has been extensively studied (1,3,4,5). The hybrid cell consists of a fuel cell cathode where dissolved oxygen in the body fluids is catalytically reduced and a sacrificial anode is galvanically oxidized.

Previous studies indicated that the primary problem was electrode polarization at the cathode resulting in a loss of power (6). The development of porous palladium black cathodes using powder metallurgy techniques results in a substantial increase in the *in vivo* power density obtainable from these cells (7). This paper addresses itself to the corrosion of anode materials. A high corrosion rate for an anode material for the hybrid cell can be a distinct disadvantage because of its proposed goal as a long term power source, the increase in waste products in the host and the possible weakening of the anode with built up corrosion products. (8,9) Weight loss studies are not sufficiently sensitive to adapt to *in vivo* corrosion studies (9,10) while for in vitro they produce an average corrosion rate unsuitable for studying changes in corrosion rate with time. (11, 12)

A technique that can be used for both *in vitro* and *in vivo* corrosion studies of anodes is the linear polarization method developed by Stern (11, 12, 13) and used by Greene et al (9,10) to study corrosion of surgical implants. The hybrid cell draws 25μ amps/cm^2 when driving a commercial pacemaker, thus in order to obtain galvanostatic and potentiostatic anodic and cathodic polarization curves under load conditions anodes were loaded to draw approximately 25μ amp/cm^2. After loading for various periods of time, i.e., for a number of days, polarization, galvanostatic and potentiostatic curves were taken on removal of the load. Besides corrosion rates being determined by linear polarization techniques rates were estimated similar to Tafel extrapolation by drawing tangents to the polarization curves in the loaded current range. The intersection of these "Tafel" lines with the horizontally drawn line representing $E_{corrosion}$ was taken as $I_{corrosion}$.

EXPERIMENTAL PROCEDURES

Linear Polarization Technique for Determining Corrosion Rates. (9-13). This method consists of determining the Tafel slopes; β_a and β_c for both anodic and cathodic processes, and the polarization resistance, R_{pol}, of the corroding electrode under study. The polarization resistance is the slope of the linear portion of the polarization curve obtained by plotting overpotential versus current density for the first 10 to 20 millivolts of overpotential of the electrode.

The values of β_a and β_c and R_{pol} are then used to calculate the corrosion rate expressed as a corrosion current I_{corr} from the following equation

$$I_{corr} = \frac{\beta_a \beta_c}{2.3\ R_{pol}(\beta_a + \beta_c)} \qquad (1)$$

A Tacussel PIT 20 - 2X potentiostat and a polarization cell fabricated according to Greene (14) was used for in vitro studies. In determining R_{pol} and for galvanostatic measurements a constant current source, a Keithly Model 225 was used for accurate low currents. A Lauda Circulator Model K-2 kept samples at a body temperature of 37 $\pm$ .1 C.

In in vivo studies the current was passed through a test electrode and a porous palladium black electrode which is implanted with the test electrodes for use as the auxiliary or counter electrode. The electrical connection to all the electrodes was made by connecting very fine Teflon-coated stainless steel wires to the electrodes and then bringing the other end of the wire out through the skin. The wire connection is insulated by a fluid impermeable neoprene coating covered with Ciba Aradite 6010 epoxy and finally Dow Corning Medical Adhesive Silicon Type A.

The potential of each test electrode was measured with respect to a silver-silver chloride electrode (standard 16 mm Beckman biopotential skin electrode) which was placed on the skin with conductive paste. All samples were 1 or 3 cm^2 in geometric area and were implanted subcutaneously in the lateral abdomen and thorax of dogs.

Solid aluminum and zinc were obtained from A. D. Mackay Co. with the following purity: aluminum, 99.99%; zinc, 99.9%. The magnesium alloy AZ31B consisted of magnesium with 3% aluminum and 1% zinc.

RESULTS AND DISCUSSION

In Vitro Measurements

Although the anodic and cathodic Tafel slopes varied over a wide range of values the I_{corr} can be shown to be approximated by

$$I_{corr} = \frac{0.026}{R_{pol}} \quad . \text{ (10)} \qquad (2)$$

Fig. 1 shows plots of linear polarization curves for determining R_{pol} for the various metals under in

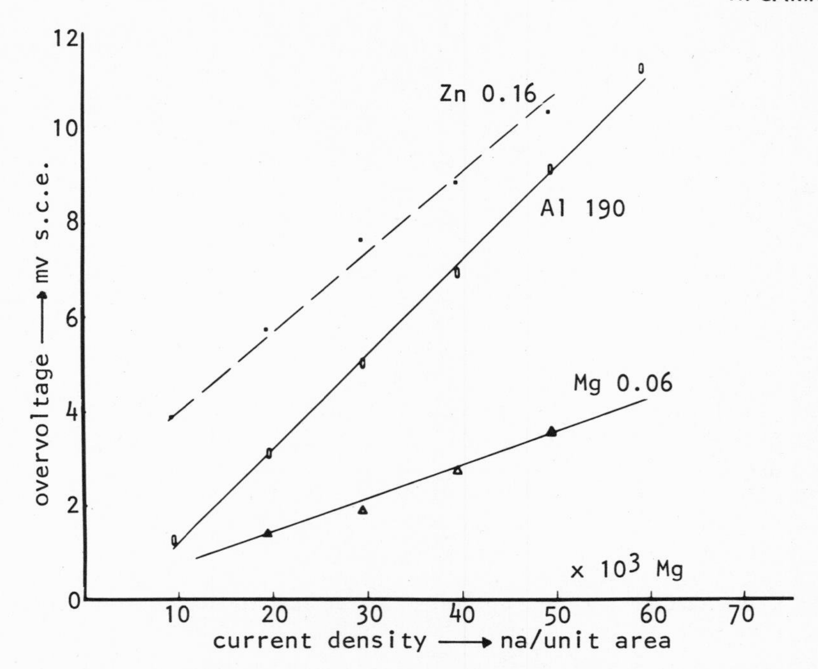

Fig. 1. In Vitro Anodic Linear Polarization Curves. For determining the R_{pol} in kiloohms per sq. cm. (under no load) for Al, Zn, Mg alloy, AZ31B, in 0.9% aqueous NaCl at 37°. Unit area = 1 sq. cm. Open to atmosphere.

TABLE I

Comparison of $I_{corrosion}$ of Anodic Materials under No Load Conditions, 0.9% NaCl, pH 6.5 to 7.0, T = 37°C.

A N O D E	I N V I T R O		
	R_{pol} ohms X 10^3	I_{corr} Equation 2 μa/sq.cm	I_{corr} Tafel line Intersection μa/sq.cm
Solid Aluminum	100 - 330	0.09 - 0.26	0.2
Solid Zinc	0.14 - 0.35	52 - 186	60
Solid Magnesium Alloy AZ31B	0.066 - 0.242	107 - 394	400 - 500

vitro conditions. Table I is a summation of *in vitro* data. It is noted that there is a range of values of R_{pol}. The intersection of the Tafel line with E_{corr} in Graph 2 is at 0.2 which checks reasonably well with the values obtained from Equation 2.

In Vivo Corrosion Measurements

The *in vivo* corrosion data under no load and loaded conditions is given in Tables II and III. Before implantations the electrodes were sterilized in an autoclave where a heavy oxide coating was deposited. Thus under no load conditions the corrosion rate particularly for aluminum was greatly reduced because of the oxide film, (15-18) while under load conditions the oxide coating appears to rupture, thereby accelerating corrosion.

It is noted from Table I that the *in vitro* corrosion rates of Zn and Mg Alloy AZ31B are to 2 to 3 orders to magnitude greater than Al. On comparing the *in vitro* data of Table I with the *in vivo* data of Table II that *in vivo* corrosion rates are reduced by an order of magnitude over the *in vitro* although the relative rates between the metals remain approximately the same. Table III illustrates that loading the anode increases the corrosion rate particularly for Al. The *in vitro* electrodes unlike the *in vivo* electrodes were not autoclaved so that they did not obtain an oxide film. Attempts to make *in vitro* measurements on autoclaved electrodes produced non reproducible and unstable data. The above results indicate that oxide film at the interface is being partially ruptured at the interface under loaded conditions (15-18) increasing the corrosion rate. As expected the disturbance to the corrosion rate of passing a current through the aluminum interface due to its more pronounced oxide coating is greater than that of zinc. *In vivo* data on the magnesium alloy under load conditions was unstable on our initial attempts. Further polarization measurements are planned in order to characterize magnesium under operating conditions.

The intersection of the Tafel line of Fig. 2 with the E_{corr} line, which in this case is 1.2 volts gave the I_{corr} recorded in Table I.

TABLE II

Comparison of $I_{Corrosion}$ of Anodic Materials Under No Load Conditions For Implantations in Dogs.

A N O D E	I N V I V O U N L O A D E D		
	R_{pol} ohms X 10^3	I_{corr} Eq.2 μa/sq.cm.	I_{corr} "Tafel"line Extrapolation μa/sq.cm.
Solid Aluminum	400 - 2000	.013 - .065	0.02
Solid Zinc	6 - 18	1.4 - 4.3	2.3
Solid Magnesium Alloy AZ31B	3 - 4	6.5 - 8.7	12

TABLE III

Comparison of $I_{Corrosion}$ of Anodic Materials Under an Approximate Loading of 25 μa/sq.cm^2.

A N O D E	I N V I V O L O A D E D		
	R_{pol} ohms X 10^3	I_{corr} Eq.2 μa/sq.cm.	$I_{corr.}$ "Tafel Line Extrapolation μa/sq. cm.
Solid Aluminum	114	0.228	0.5
Solid Zinc	5.8	4.4	5.5

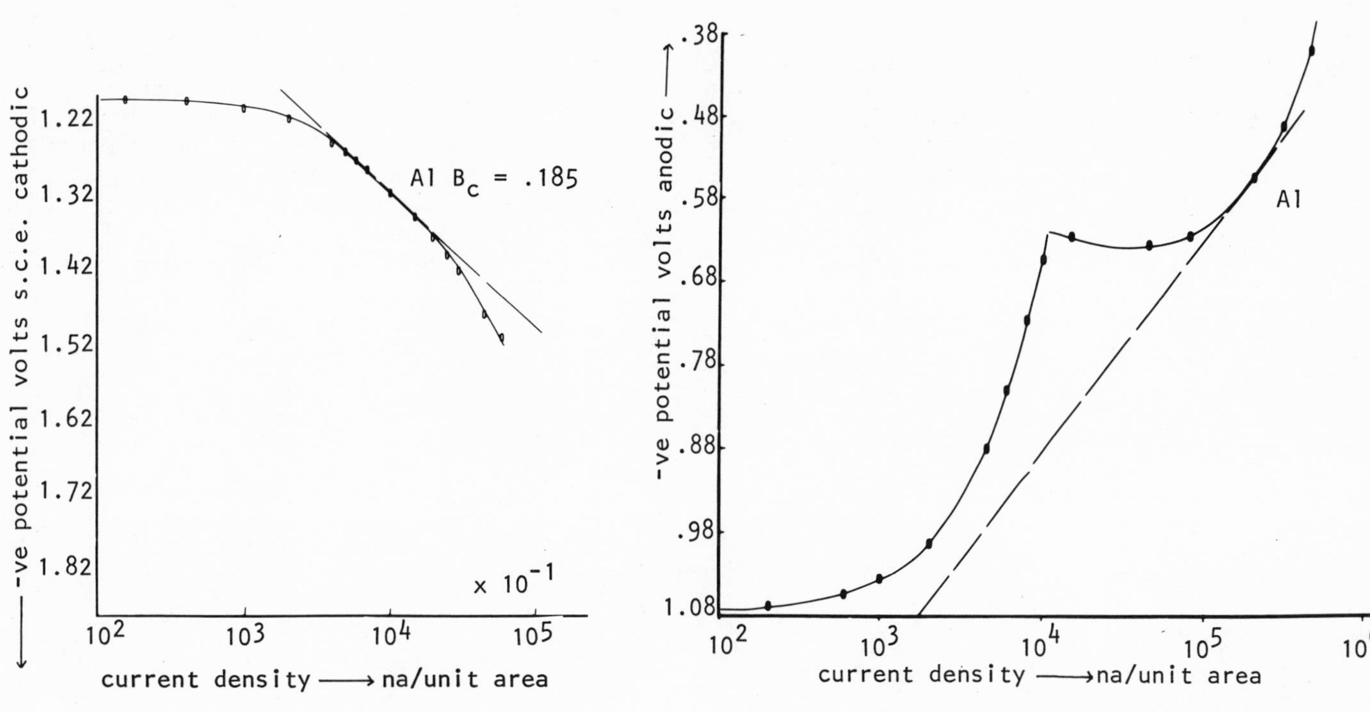

Fig. 2. In Vitro Galvanostatic Cathodic Polarization Curve for Solid Al. In 0.9% aqueous NaCl at 37°C. Unit area = 1 sq. cm. Under no load conditions. Open to atmosphere.

Fig. 3. In Vivo Galvanostatic Anodic Polarization Curve for Solid Al Implanted in a Dog. Unit area = 3 sq. cm. Loaded approximately 25 µamp/sq. cm. for 4 days. Potential vs. Ag/AgCl.

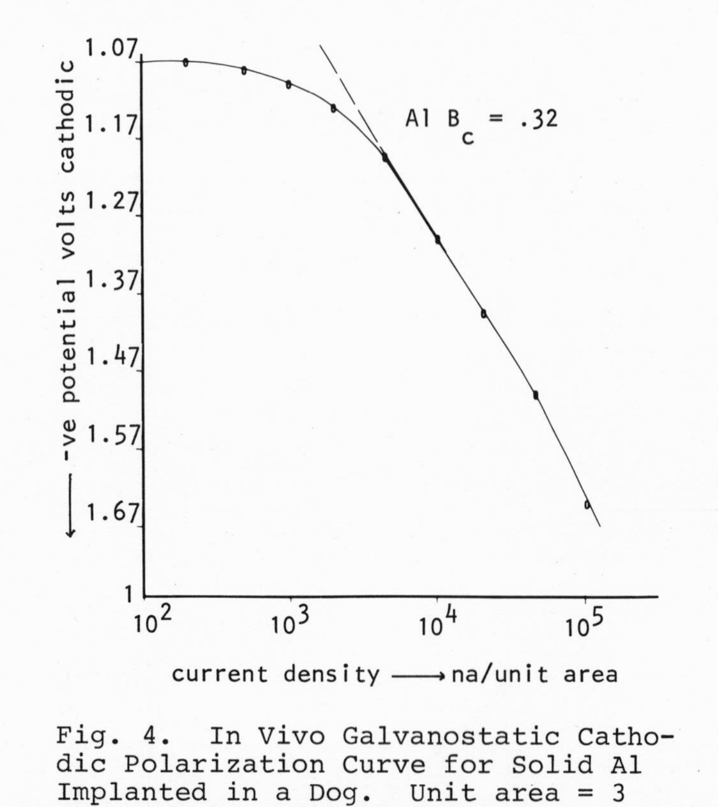

Fig. 4. In Vivo Galvanostatic Cathodic Polarization Curve for Solid Al Implanted in a Dog. Unit area = 3 sq. cm. Loaded approximately 25 μamp/sq. cm. for 4 days. Potential vs. Ag/AgCl.

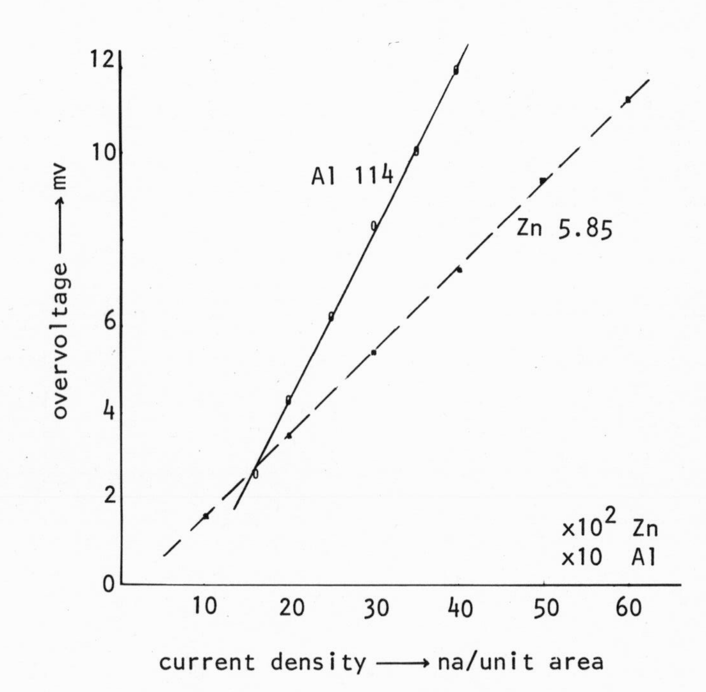

Fig. 5. In Vivo Anodic Linear Polarization Curves for Determining R_{pol} in kiloohms/sq. cm. Electrodes loaded approximately 25 μamp/sq. cm. for 4 days. Potential vs. Ag/AgCl. Unit area = 3 sq. cm.

Fig. 3 illustrates the effects of the oxide coating on aluminum with the curve at low current densities being characteristic of the corrosion rate with a substantial oxide coating. The surface appears to be considerably different under loaded current densities of 25μa/cm^2. A tangent on "Tafel" line was drawn for the curve in the operating range. The I_{corr} determined from the intersection of the tangent lines with the horizontal E_{corr} line for the anodic and cathodic galvanostatic curves check closely.

Fig. 5 is an illustration of linear polarization data under loaded *in vivo* conditions. Although the slopes of the curves vary on a dog, from measurement to measurement and from dog to dog the data points line up well for a given measurement.

The present pacemaker system we are using requires 610μa continuously when operating from a cell voltage of 0.5 volts. Over a ten year period the faradaic weight loss in an anode of trivalent Al needed to supply the above current is 17.6 gms. The weight loss based on I_{corr} measurements under the above load is 0.17 gms. or about 1.0% of the faradaic loss. Divalent Zn on the other hand because of a 300 millivolt higher voltage would need a current of about 400μa or a faradaic weight loss of 40 gms. over a ten year period. The weight loss based on I_{corr} for Zn of an equivalent geometrical area is 11.4 gms. or 28% of the faradaic. Divalent Mg with its 600 millivolt higher voltage requires a current of approximately 270μa or a faradaic weight loss of 10.7 gms. The weight loss based on I_{corr} for Mg of an equivalent geometrical area under no load is 8 gms. or 75% of the faradaic weight loss. After operating *in vivo* under load the Mg galvanostatic and potentiostatic curves were too unstable for corrosion data. The *in vivo* Mg corrosion appears to be much greater under load than no load.

The addition of protein BSA to the 0.9% NaCl solution in a pH range of 5-6 cut in half the *in vitro* corrosion rates of Zn. Organic substances have been shown to strongly effect metallic dissolution rates (19).

The above comparatively low corrosion rates and histological studies for Al indicate that Al, with an efficient catalytic porous cathode is a good combination for an implantable power cell.

ACKNOWLEDGEMENTS

This study was made possible by the support of the Biomaterials Program of NSF Grant #GH33748.

REFERENCES

1. Schaldach, M. Von, "An Electrochemically Actuated Pacemaker," Wiederelebung and Organersaty, Vol. 5, No. 2, pp. 75-84.
2. Dack, S. and Donaso, E., "Heart Block with Stokes-Adams Syndrome Indicator and Results of Cardiac Pacing in Part I of Advances in Cardiac Pacemakers, S. Furman, Ed., Annals of N.Y.A.S., Vol. 167, 1969.
3. Racine, P., "Power Generation from Implantable Electrodes," M.S. Thesis in Biomedical Engineering, Drexel University, 1966.
4. Roy, O.Z., "Biological Energy Sources: A Review," Bio-Medical Engr., Vol. 6, No. 6, June 1971, pp. 250-256.
5. Tseung, A.C.C., King, W. J. and Wan, X.C., "An Encapsulated, Implantable Metal-Oxygen Cell as A Long-Term Power Source for Medical and Biological Applications," Med. Biol. Eng., Vol. 9, pp. 175-184, 1971.
6. DeRosa, J.F., Beard, R.B., and Hahn, A. W. "Fabrication and Evaluation of Cathode and Anode Materials for Implantable Hybrid Cells," IEEE Trans. Bio-Med. Eng., Vol. BME-17, pp. 324-330, Oct. 1970.
7. Beard, R.B., DeRosa, J.F., Koerner, R.M., Dubin, S.E., and Lee, K.J., "Porous Cathodes for Implantable Hybrid Cells," IEEE Trans. Bio-Med Eng., Vol. BME-19, May 1972, pp. 233-238.
8. Payer, J.H. and Staehle, R.W. "Localized Attack on Metal Surfaces," p. 211 NACE-2 Corrosion Fatigue Edited by Devereux, O.E., McEvily, A.J., and Staehle, R.W., Nat.Assoc. of Corrosion Eng., 1972.
9. Colangelo, V.J., Greene, N.D., Kettlekamp, D.B., Alexander, H. and Campbell, C.J., J. Bio-Med Mater. Res., Vol. 1, 1967.
10. Greene, N.D., and Jones, D.A., "Corrosion of Surgical Implants." J. of Materials, Vol. 1, No. 2, June 1966, pp. 345-353.
11. Stern, M. and Weisert, E.D., "Experimental Observations on the Relation Between Polarization Resistance and Corrosion Rate," Proc. Am. Soc. Test Mat., Vol. 59, 1959, p. 1280.

12. Stern, M. and Geary, A.L., "Electrochemical Polarization I. A Theoretical Analysis of the Shape of Polarization Curves," J. Electrochem. Soc., Vol. 104, p. 56, 1957.
13. Stern, M. "A Method For Determining Corrosion Rates From Linear Polarization Data," Corrosion, Vol. 14, Sept. 1968, pp. 60-64.
14. Greene, N.D., "Experimental Electrode Kinetics," Rensellaer Polytechnic Institute Publications, N.Y., 1965.
15. Plumb, Robert C. "The Aluminum Anode" Alco Research Laboratories, New Kensington, Pa., Report No. 8-58-4.
16. Levis, J. E. and Plumb, R. C., "Studies of the Anodic Behavior of Aluminum" Journal of the Electrochem. Soc., Vol. 105, No. 9, p. 496, 1958.
17. Plumb, R.C., "Studies of the Anodic Behavior of Aluminum II & III" J. of Electrochem. Soc., Vol. 105, No. 9, p. 498, 1958.
18. Van Rysselberghe, Pierre, "Electrochemical Kinetics of the Anodic Formation of Oxide Films," J. of Electrochem. Soc., Vol. 6, No. 4, p. 355, 1959.
19. Svare, C.W., "Anodic Dissolution Studies of Selected Metals In A Simulated In Vivo Environment," Dissertation in Metallurgy and Materials Science, E. Korostoff, Advisor, V.P., 1969.

Index